Springer Series in Optical Sciences

Volume 175

Founded by
H. K. V. Lotsch

For further volumes:
http://www.springer.com/series/624

Springer Series in Optical Sciences

The Springer Series in Optical Sciences, under the leadership of Editor-in-Chief William T. Rhodes, Georgia Institute of Technology, USA, provides an expanding selection of research monographs in all major areas of optics: lasers and quantum optics, ultrafast phenomena, optical spectroscopy techniques, optoelectronics, quantum information, information optics, applied laser technology, industrial applications, and other topics of contemporary interest.

With this broad coverage of topics, the series is of use to all research scientists and engineers who need up-to-date reference books.

The editors encourage prospective authors to correspond with them in advance of submitting a manuscript. Submission of manuscripts should be made to the Editor-in-Chief or one of the Editors. See also www.springer.com/series/624

Sébastien Forget · Sébastien Chénais

Organic Solid-State Lasers

Springer

Sébastien Forget
Sébastien Chénais
Laboratoire de Physique des Lasers
Université Paris 13, Sorbonne Paris Cité, CNRS
Villetaneuse
France

ISSN 0342-4111 ISSN 1556-1534 (electronic)
ISBN 978-3-642-36704-5 ISBN 978-3-642-36705-2 (eBook)
DOI 10.1007/978-3-642-36705-2
Springer Heidelberg New York Dordrecht London

Library of Congress Control Number: 2013938168

Printed on acid-free paper

Springer is part of Springer Science+Business Media (www.springer.com)

Preface

People working in the world of organic photonics and organic electronics are used to hearing the following statement: a prerequisite to success in this field is to acquire interdisciplinary knowledge "at the crossroads" between at least two or three domains: chemistry, material science, electronics, and photonics. Working on organic lasers is no exception to this rule, and requires at the same time solid foundations in laser physics and photonics, mixed with a good knowledge of photophysics of organic pi-conjugated systems. This is probably what makes this field so exciting and collaborations so fruitful. However, this also raises some difficulties in transmitting knowledge and know-how to students or newcomers in the field having totally different backgrounds.

The authors have experienced this challenge, as they started in the field a few years ago, coming with a previous knowledge in laser physics and photonics, but no experience with organics. They wrote this book with the following guideline in mind: gathering in the same place introductory material, designed to give to a reader with no particular background in organic photophysics the keys to understand the challenges in the field, together with a state of the art, at the time of writing, of the recent advances and trends in organic lasers.

Why Organic Solid-State Lasers?

The story of organic lasers is almost as long as the history of lasers themselves: while the very first laser came to birth in 1960 [1], the first organic laser followed soon after, in 1966 [2, 3]. Organic lasers were at that time known as liquid "dye" lasers, a term inherited from the textile industry, and based on solutions of π-conjugated highly luminescent molecules. This kind of sources marked a true revolution in laser science, as the first widely tunable laser sources [4], leading to impressive progresses in spectroscopy in the following years. Additionally, the unique broad spectrum of organic dyes, coupled to the possibility of continuous-wave operation demonstrated in 1970 [5] opened the door to the first ultrashort (femtosecond) laser [6, 7]. Although often considered now far from being "user-friendly", liquid dye laser are still today popular in laboratories as sources of tunable visible radiation. However, their cumbersome designs, their liquid state, and the inconvenience linked to toxic solvents and dyes prevent them from being used for "real life" applications. Very early (as soon as 1967) it was then proposed to incorporate dyes in solid-state polymeric matrices [4] which appeared to be a promising route to build broadly tunable sources that would have the benefit to be compact, convenient, and manufactured at low costs. However, organic solid-state lasers did not manage, from then and up to now, to enter the marketplace: the main reason is probably that there is a fundamental economic contradiction between the inherent bad photostability of organic molecules and the high cost of the pump source, in general a pulsed laser. Subsequent efforts then naturally went in the direction of improving the dye and host matrix photostability, and/or decreasing the required pump threshold intensity so that alternatives to pulsed lasers may be found.

The field of organic solid-state lasers experienced a second birth at the end of the 1990s, following the discovery of organic semiconductors (mainly driven by the potentially huge market foreseen for Organic Light-Emitting Diodes [8]), and progresses came both from material science and deposition process technology. Very low thresholds could be demonstrated in thin-film based organic semiconductor lasers; furthermore, the devices turned out to be easy-to-handle and compact, while keeping all the advantages of organic materials [9, 10]. The question was then whether an *organic laser diode* would be ever realized, that is a device pumped with an injected current rather than with an optical source, in the same way as inorganic semiconductor lasers work. This issue drove considerable efforts and is still now a major inspiration railroad for organic laser research, although no demonstration of such a device has been reported yet.

The first two chapters are written in a "back-to-basics" spirit, with in-depth (although non exhaustive) description of the physics useful for entering the pluridisciplinary field of organic lasers. The next chapters deal with very recent research, reviewing state of the art and last trends in the field (for the interested reader, complementary information can also be found in some recently published review papers [11–13]). The book is organized as follows:

Chapter 1 is dedicated to the reader not familiar with lasers physics. It proposes a quick and simple picture of the main laser principles, but is not intended to explain laser physics in depth, as this task is performed in an impressive number of excellent teaching books [14, 15].

In Chap. 2, we introduce the specificities linked to organic materials as lasing media. To this aim, we first remind some photophysics basics that are useful to understand the following, before detailing the various aspects important for organic lasers and specific to them, and which are therefore not usually treated in classical textbooks on lasers. A first question that we may try to answer in this chapter is the following: what are the key features that make organic solid-state lasers (OSSL) different from other types of lasers, like those based on inorganic semiconductors, rare-earth doped crystals, or gas mixtures, to name a few? This chapter has been written with the intention to present organic laser physics and relevant photophysics at a basic introductory level, hoping that it will be accessible for a reader with a background in general physics and laser physics but with no specific knowledge in photophysics.

Chapter 3 is devoted to organic materials, both dyes and organic semiconductors, from the description of their chemical structure and emission properties (and the way to measure them, especially optical gain) until the description of the fabrication techniques mainly used to make devices out of them.

In Chap. 4, the focus is set on the lasing architectures. Organic materials can be shaped and processed in such various ways that almost every resonator configuration can be adapted to them. Recent results for each cavity type are reviewed, with special emphasis on periodically structured resonators and vertical external cavity surface-emitting lasers.

Chapter 5 is a review of the last advances in the field of organic solid-state lasers, including the quest for electrically driven organic lasers, indirect pumping solution featuring inorganic diodes, as well as promising results in the field of nanoplasmonics where organic materials have an interesting role to play.

Finally, Chap. 6, the last chapter deals with the rapidly growing applications of organic lasers. The natural playground associated to organic lasers is spectroscopy, where the modest output power is not problematic and broad wavelength agility is a strong advantage. The simplicity of the fabrication process now makes organic sources attractive for integration onto miniature spectroscopic systems. More recently, organic lasers have been also used for chemical sensing, for example to detect trinitrotoluene (TNT) using conjugated polymer films [16]. The easy integration of organic solid-state lasers makes then ideal sources for lab-on-a chip sensors for biophotonics, coupled to microfluidic devices [17]. Organic semiconductor lasers may also find their place in data communications, for short-haul data transfer or optical amplifiers for example.

We are very grateful to people—students, post-doctorants, and technical staff—who contributed to the development of the organic laser research activity in the Lasers Physics Laboratory at Paris 13 University, and hence contributed to make this book possible. We would like to express special thanks to Hadi Rabbani-Haghighi as our first Ph.D. student on the topic, and also to Mélanie Lebental, Oussama Mhibik, Iryna Gozhyk, and Tatiana Leang. We finally gratefully thank Marie-Claude Castex, Elena Ishow, Alain Siove, Dominique Adès, and Laurent Vernac for fruitful discussions and support which helped in writing this book.

Villetaneuse

Sébastien Forget
Sébastien Chénais

References

1. T.H. Maiman, Nature. **187**, 493 (1960)
2. P.P. Sorokin, R. Lankard, Stimulated emission observed from an organic dye, chloroaluminium phtalocyanine. IBM J. Res. Develop. **10**, 162–163 (1966)

3. F.P. Schafer, W. Schmidt, J. Volze, Organic dye solution laser. Appl. Phys. Lett. **9**(8), 306–309 (1966)
4. B.H. Soffer, B.B. McFarland, Continuously tunable, narrow-band organic dye lasers. Appl. Phys. Lett. **10**(10), 266 (1967)
5. O.G. Peterson, S.A. Tuccio, B.B. Snavely, {CW} operation of an organic dye solution laser. Appl. Phys. Lett. **17**(6), 245–7 (1970)
6. E.P. Ippen, C.V. Shank, A. Dienes, Passive mode locking of the cw dye laser. Appl. Phys. Lett. **21**(8), 348–350 (1972)
7. R.L. Fork, B.I. Greene, C.V. Shank, Generation of optical pulses shorter than 0.1 psec by colliding pulse mode locking. Appl. Phys. Lett. **38**(9), 671–672 (1981)
8. C.W. Tang, S.A. Vanslyke, Organic electroluminescent diodes. Appl. Phys. Lett. **51**(12), 913–915 (1987)
9. N. Tessler, G.J. Denton, R.H. Friend, Lasing from conjugated-polymer microcavities. Nature. **382**(6593), 695–697 (1996)
10. F. Hide et al., Semiconducting polymers: A new class of solid-state laser materials. Science. **273**(5283), 1833–1836 (1996)
11. I.D.W. Samuel, G.A. Turnbull, Organic semiconductor lasers. Chem. Rev. **107**(4), 1272–1295 (2007)
12. S. Chénais, S. Forget, Recent advances in solid-state organic lasers. Polym. Int. **61**(3), 390–406 (2012)
13. C. Givras, M. Pollnau, Organic solid-state integrated amplifiers and lasers. Laser and Photonics Revi. **6**(4), 419–462 (2012)
14. A.E. Siegman, Lasers. (1986)
15. O. Svelto, *Principles of Lasers*. (Macmillan Publishers Limited, New York, 1998)
16. J.S. Yang, T.M. Swager, Fluorescent porous polymer films as TNT chemosensors: Electronic and structural effects. J. Am. Chem. Soc. **120**(46), 11864–11873 (1998)
17. C. Vannahme et al., Plastic lab-on-a-chip for fluorescence excitation with integrated organic semiconductor lasers. Opt. Express. **19**(9), 8179 (2011)

Contents

Chapter 1
Laser Basics

Abstract The word laser is the acronym for Light Amplification by Stimulated Emission of Radiation. Lasers are devices that produce or amplify a beam of usually narrow, low-divergence light with a well-defined wavelength within the optical region of the electromagnetic spectrum, covering the infrared, the visible and the ultraviolet. This chapter is not intended to explain laser physics in depth (this task is performed in an impressive number of excellent teaching books [1]), but to set out some general ideas on lasers as a reminder or an introduction for people coming from other fields. Some of the concepts explained here will be used in the following chapters of this book, with a sharpened point of view limited to organic materials.

1.1 What is a Laser?

A laser, in the literal sense of the word, is an *amplifier* of optical radiation, and consists basically of two fundamental elements:

- An amplifying medium (often referred to as the "gain medium"), which can be a solid, a liquid or a gas. The intensity of a given light wave will be amplified after passing through this medium, if the energy of the corresponding photons matches the energy levels of the atoms, molecules or ions forming the gain medium. The physical associated process is called *stimulated emission.*
- A system to excite the amplifying medium (also called a pumping system). This creates the conditions for light amplification by supplying the necessary energy. There are different kinds of pumping systems: optical (e.g. flash lamps, arc lamps or tungsten-filament lamps, diodes or other lasers), electrical (gas discharge tubes, electric current in semi-conductors) or even chemical.

These two components together form a laser amplifier. However, in the usual sense of the word, a laser also incorporates an optical resonator (or cavity) to

S. Forget and S. Chénais, *Organic Solid-State Lasers*,
Springer Series in Optical Sciences 175, DOI: 10.1007/978-3-642-36705-2_1,

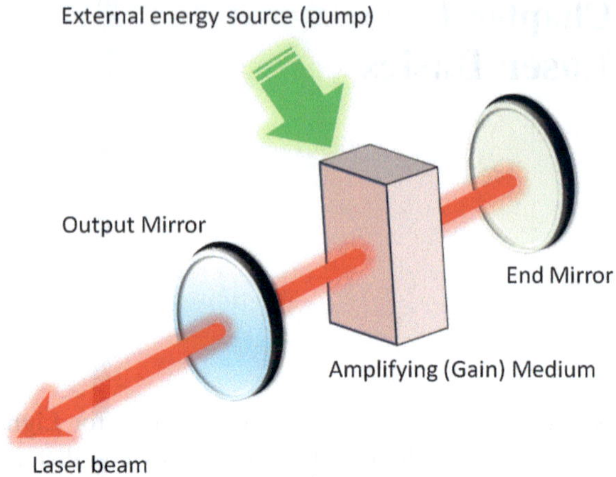

Fig. 1.1 General scheme of the different laser building blocks: an amplifying medium, a pump source and a cavity

enhance the amplification and shape the output spectrum and beam geometry. The different components that make up a basic laser are illustrated in the diagram below (Fig. 1.1).

1.2 Light–Matter Interaction

Light can basically interact with matter following three processes, depicted in Fig. 1.2 for a simplified two-level system. Here, the matter is composed of optically active elements in "solution" in a gas, plasma, solid or liquid medium. These elements can be atoms, ions, molecules, free radicals or electrons (for simplicity, named as "atoms" in the following). Their energy levels are quantified and are such as light of a certain frequency can interact with the population found in these levels. More precisely, let us consider two energy levels E_1 and E_2 ($E_1 < E_2$) and some light with frequency $h\nu = E_2 - E_1$.

1. Absorption: An atom in a lower level absorbs a photon of frequency $h\nu$ and moves to an upper level.
2. Spontaneous emission: An atom in an upper level can decay spontaneously to the lower level and emit a photon of frequency $h\nu$ if the transition between E_2 and E_1 is radiative. This photon has a random direction and phase.
3. Stimulated emission: An incident photon causes an upper level atom to decay, emitting a "stimulated" photon whose properties are identical to those of the incident photon. The term "stimulated" underlines the fact that this kind of radiation only occurs if an incident photon is present.

For a radiative transition, these three mechanisms are always present at the same time. To make a laser medium, conditions have to be found that favor stimulated emission over absorption and spontaneous emission. Thus, both the right medium and the right conditions must be chosen to produce the laser effect.

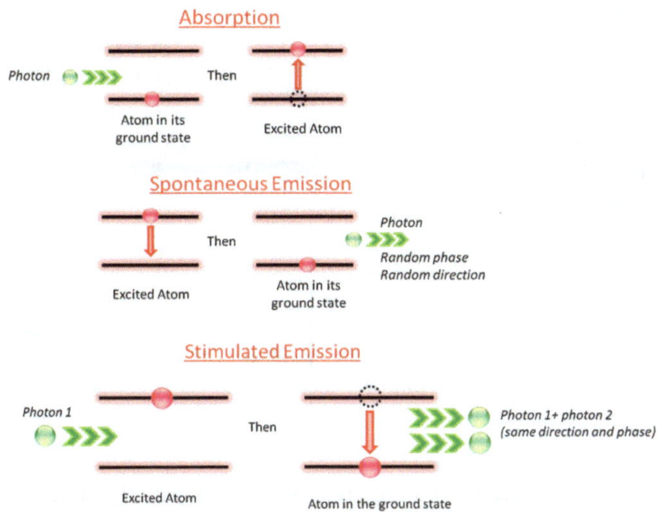

Fig. 1.2 Fundamental processes of light-matter interaction

Einstein rules state that a photon of energy $h\nu$ incident on a model 2-level atom has an equal probability to be absorbed by a ground-state atom as to be duplicated (or "amplified") through stimulated emission by an excited-state atom (absorption and stimulated emission have the same cross sections). Absorption and stimulated emission are two reciprocal processes that have the same probability. To favor stimulated emission over absorption for a given transition, the number of atoms in the excited upper state N_2 must be higher than the number of atoms N_1 in the lower state of the optical transition. This is called population inversion, and can be expressed by

$$\Delta N = N_2 - N_1 > 0.$$

1.3 Population Inversion

At thermodynamic equilibrium, the distribution of the atoms between levels is given by Boltzmann's Law:

$$N_2 = N_1 \cdot \exp\left[\frac{-(E_2 - E_1)}{kT}\right]$$

where k is the Boltzmann constant and T the temperature.

In this case, N_2 is always less than N_1 and no light amplification can occur. A laser is then necessarily an "out-of-equilibrium" system, and this state can be reached through external energy addition via a process known as "pumping" in order to raise enough atoms to the upper level (Fig. 1.3). The pumping process can be performed through electrical, optical or chemical means.

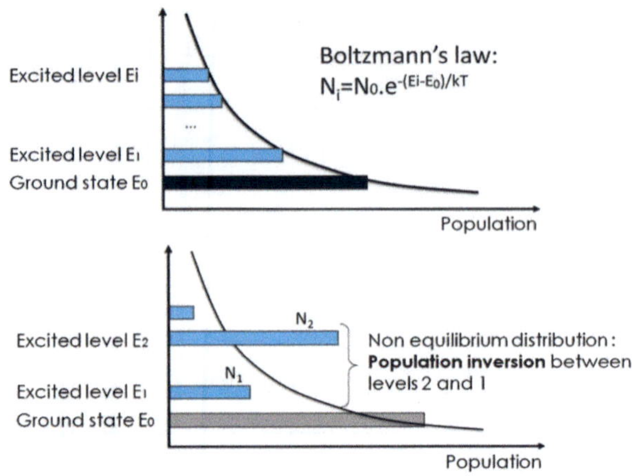

Fig. 1.3 Under thermal equilibrium (*up*): energy levels are populated according to the Boltzmann distribution. Under pumping (*bottom*), the system is set out-of-equilibrium and population inversion can be achieved

Atomic transitions have both a radiative and a non-radiative part, and not all of them are good candidates to create population inversion. The upper level can be depleted through both non-radiative decay and spontaneous emission (also called fluorescence). Storing atoms in the excited level could be challenging if its fluorescence lifetime is very short or if the non-radiative paths are efficient. The lower level of the lasing transition has also to be considered: choosing a lower level which is not the ground state will greatly limit the population N_1, which may even be zero if E_1 is several times kT above the ground state. In this case only one atom excited to the E_2 level is enough to create a population inversion between E_2 and E_1.

In optical pumping, at least three different energy levels are needed to create a population inversion. Figure 1.4 (left) illustrates such a so-called three-level system. It shows the pumping transition (between E_1 and E_3) and the laser transition (between E_2 and E_1). The goal is to store atoms in level E_2 by absorbing "pumping" radiation. Ideally, the excited atoms must quickly decay from level 3 to level 2, where they remain much longer than the time it took to go from 3 to 2. This time is the excited state lifetime of this level. The decay by stimulated emission to the ground state (level E_1) is the lasing transition. Despite its simplicity, this system is somewhat inefficient because the ground state has a large population at thermodynamic equilibrium and at least half of this huge population must be excited to simply reach population inversion. Surprisingly, for historical reasons the first laser ever made was of this type and used a ruby (Cr^{3+}:Al_2O_3), that is an aluminium oxide crystal matrix doped with chromium ions (Cr^{3+}) whose energy levels are suited to create the laser effect.

A much more efficient laser scheme is the four-level system (Fig. 1.4, right). Here, the pumping transition (optical pumping) and the laser transition occur over a pair of distinct levels (E_0 to E_3 for the pump and E_1 to E_2 for the laser). E_1 is chosen to be sufficiently high in energy relative to the ground state E_0 so that the thermal population at thermodynamic equilibrium is negligible. Similarly, atoms in level 3 or level 1 decay very rapidly towards levels 2 and 0, respectively. Unlike

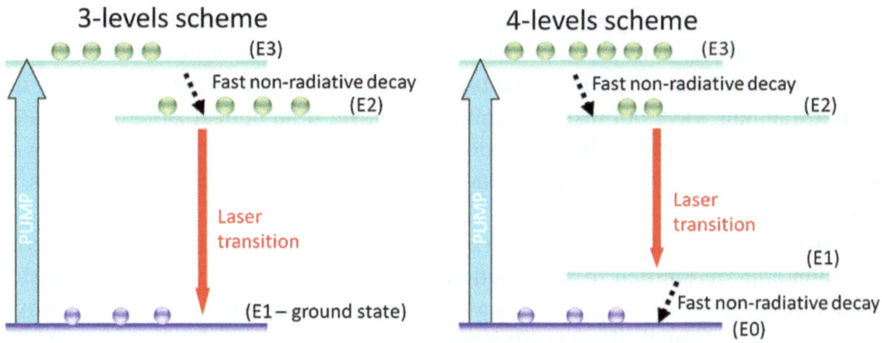

Fig. 1.4 Example of three-level (with terminal laser level on the ground state here) and four-level systems

the three-level system, as soon as a few atoms (just one in an ideal case) is promoted to level 2, population inversion occurs. One of the best known medium operating in this way is neodymium Nd in various matrices—the most famous being YAG (Nd^{3+}:$Y_3Al_5O_{12}$).

1.4 From an Amplifier to an Oscillator

1.4.1 Role of the Resonator

The previous section showed how population inversion can be set by choosing adequate spectroscopic systems with well-suited energy levels and decay rates. However, establishing a population inversion is not enough to claim a laser effect. As stated previously, stimulated and spontaneous emissions are competing with each other. To create conditions where stimulated emission becomes predominant over spontaneous emission, an optical cavity is generally required. The cavity or resonator is composed of several mirrors that reflect the beam back and forth through the amplifying medium (Fig. 1.5).

The simplest optical cavity is a linear cavity composed of two facing mirrors separated by a distance d. The curvature radii are R_1 and R_2. In this kind of optical

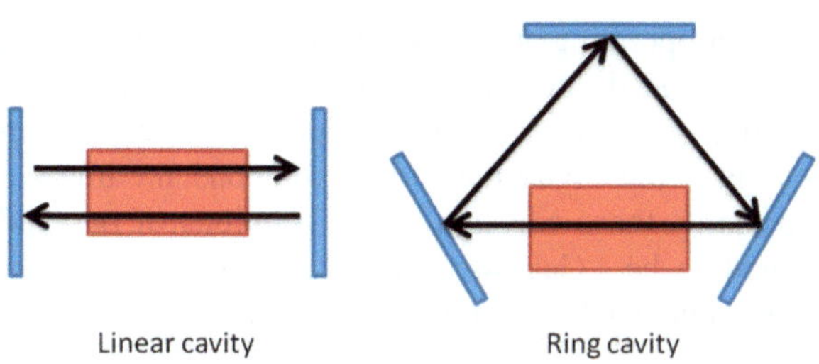

Fig. 1.5 Two examples of resonator configurations

cavity, a standing wave pattern occurs between the mirrors. Some optical elements (lenses, polarizers, active components…) could eventually be inserted inside the cavity. Another well-known cavity type is the ring cavity, where light may not form a standing wave but instead a travelling wave, if an optical isolator is set into the cavity to prevent one flow direction. In the following, we will only deal with linear cavities, but the principles and methods are the same for both types of resonators.

When the laser starts up, the first emitted photons are due to spontaneous emission and are emitted in all directions. Of course, a small part of the emission occurs along the axis of the laser cavity. These spontaneous photons can travel back and forth through the amplifying medium (in the case of a linear cavity for example). At each stage, they are amplified by stimulated emission: over time, the amount of coherent light in the cavity increases considerably. The light confinement increases the probability of stimulated emission over spontaneous emission. At the same time, the cavity acts as a filter due to the numerous round trips: only the light rays perfectly aligned with the cavity axis will stay inside the resonator, and the spatial geometry of the laser beam is then defined by the cavity properties. Moreover, certain frequencies will be favored (the resonance frequencies of the cavity): the cavity shapes the output spectrum of the laser (see later).

A laser oscillator is similar to an electronic oscillator: it requires an amplifier, a feedback element, and a gain-control element to fix the amplitude of oscillations to a stable value (we mention here only oscillators working in the continuous-wave regime). The optical gain medium is the amplifier, the cavity serves as the feedback element and sets the cavity resonant frequencies (see Sect. 1.4.3), and the role of the gain-controlling element is here played by gain saturation [1] in the medium. Gain saturation means that the available optical gain (related to the amount of population inversion) decreases with the laser intensity, because every photon produced by stimulated emission "consumes" one atom in the excited state, which tends to lower the population inversion and hence the gain. As a result for a given value of stationary inversion population, there will be only one possible value of intensity that will enable a sustained stable operation, or in other words the system is self-stable (see Fig. 1.6).

1.4.2 Spatial Characteristics of the Emitted Laser Beam

A laser operating at steady state produces a light wave whose spatial structure does not change despite numerous round trips inside the cavity. This means that the spatial properties (wavefront curvature, beam extension…) of the wave propagating in the cavity have to remain constant after each round trip. Mathematically speaking, these waves must be solutions of paraxial Helmholtz equation and are found to be "Gaussian" waves. Their intensity distribution shape in the plane perpendicular to the axis of propagation is a Gaussian curve and writes:

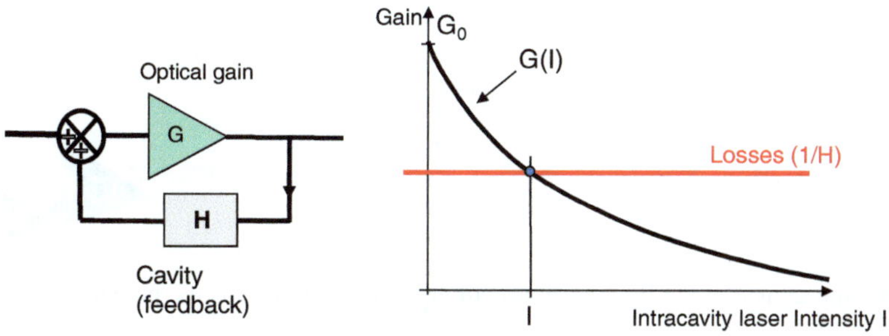

Fig. 1.6 Analogy between an electronic oscillator and an optical oscillator. The cavity determines the resonant frequencies; gain saturation (decrease of gain with intensity because stimulated emission "consumes" population inversion) is the stabilizing mechanism which allows obtaining stable oscillation amplitudes (output intensities)

$$I(r,z) = I_0(z)e^{\frac{-2r^2}{w^2(z)}}$$

where r is the radial coordinate, w(z) the beam waist defined as the radius where the intensity is equal to $1/e^2$ times its maximum value (see Fig. 1.7).

A Gaussian wave propagates in a slightly different way as a beam described by classical geometrical optics.

If the origin on the propagation axis z is chosen where the size of the beam is minimal (this position, known as the laser beam waist w_0, corresponds to the position where the radius of curvature of the wave is infinite), the expansion of the beam size with increasing z is governed by the equation given in Fig. 1.8. The beam divergence is given by the limit of w/z when z goes to infinity: $\theta = \lambda/(\pi w_0)$.

It is important to note that the beam profile defined above is one single solution of the paraxial wave equation, called the fundamental Gaussian mode. Other solutions exist and each oscillation in the resonator is a linear combination of those solutions or modes. The transverse structure of the laser beam could then become much more complex than the one seen in Fig. 1.8, exhibiting a rectangular or

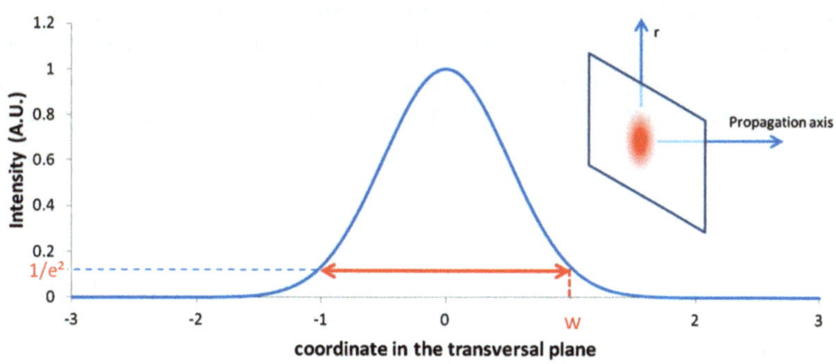

Fig. 1.7 Appearance of a Gaussian beam: distribution of intensity in a plane perpendicular to the direction of propagation

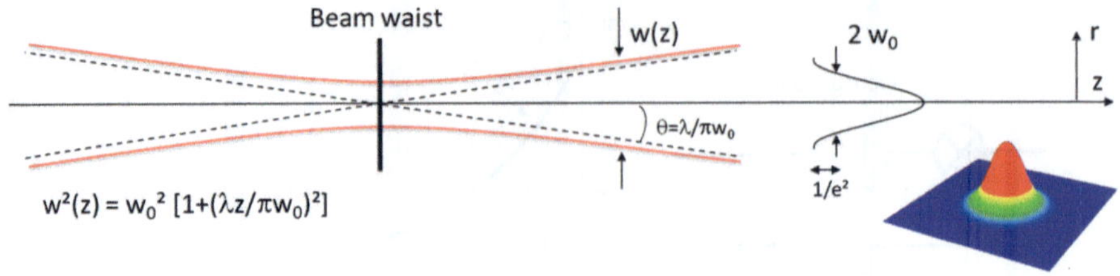

Fig. 1.8 Appearance of the light beam according to its position (z is the propagation axis)

cylindrical symmetry with an alternation of dark and brilliant fringes. For nearly all applications, those complex structures are not welcome and a fundamental Gaussian beam is highly preferred.

The laws of Gaussian optics govern the beam shape and propagation for perfect laser cavities such as the one depicted before. In most "real" cases, the cavity can be very different (including diffraction gratings, waveguides, intracavity elements with anisotropic properties, heating of the laser medium…). A parameter, named the M^2 factor, is defined to quantify the difference between a perfect expected Gaussian laser beam and the real one. In its simplest form, M^2 is the ratio between the divergence θ of the real laser beam and the divergence of a perfect Gaussian beam having the same waist:

$$M^2 = \theta \cdot \frac{\pi w_0}{\lambda}$$

Obviously, M^2 is equal to 1 for a fundamental Gaussian beam and increases when the laser beam departs from a perfect Gaussian beam. As the diffraction-limited divergence is the minimal divergence for any light beam, $M^2 > 1$.

1.4.3 The Spectrum of a Laser Oscillator

As previously stated, the cavity also sets the spectrum emitted by the laser. A linear cavity with two facing mirrors is basically a Fabry-Pérot (FP) interferometer. Consequently, only a given set of frequencies can successfully propagate in the resonator. Those frequencies, called longitudinal modes, are given by the FP formula:

$$\nu_p = \frac{pc}{2L}$$

where p is an integer, c the speed of light in vacuum and L the optical length of the cavity. p is usually large for "classical" macroscopic cavities (a few centimeters long) and the spectral interval between two consecutive modes (the free spectral range (FSR) of the FP) is very small—for example for a 10 cm-long resonator, $\Delta\lambda$ is typically in the picometer range—and it is usually difficult to observe these different modes with classical spectrometers. In contrast, microcavities have widely spaced longitudinal modes, and such a microlaser can run in a single longitudinal mode, while this is not generally the case with large cavities, where a "comb" of modes may have enough gain to oscillate simultaneously (Fig. 1.9). In theory, a purely homogeneous medium (all atoms or molecules of the active medium have the same spectrum) will have a single-mode steady-state spectrum because of mode competition: only the mode with the highest gain survives and takes all the available gain.

A laser is often described as monochromatic in the sense that only one color is visible to the naked eye. It is generally true because the spontaneous emission spectrum is narrow (a fraction of nanometers)—however it has to be reminded that several wavelengths are generally emitted as stated above. But in some specific cases such as organic molecules, a much wider transition (several tens of nanometers) is observed because of the presence of numerous rotational and vibronic energy levels, forming wide energy bands instead of discrete energy levels. In those cases, the laser emission spectrum can be tunable, it may be rather broadband, and its precise shape (number and position of each mode) is governed by the geometry of the resonator.

1.5 Oscillation Condition

The condition for a laser to oscillate is remarkably simple: the gain due to the amplifying medium has to overcome the losses in the cavity.

The effective gain of an amplifying medium is simply defined by the ratio between the power (in Watts or photon/s) of the outgoing beam P_{out} and the power of the incoming beam P_{in}:$G = P_{out}/P_{in.}$

The losses are related to defects in the amplifying medium (diffusion), reabsorption, reflections on the optical elements inside the resonator (including the gain medium itself), which are generally unwanted and non controllable, and of course the "useful" losses provided by the finite reflectivities (R_1 and R_2). The latter are useful because if we set R_2 as the sole output coupler, then the fraction $1- R_2$ of the intracavity laser field power will emerge out of the cavity to make the observable laser beam.

When the laser is operating continuously, the output power is constant despite the fact that the number of photons in the cavity increases when passing through

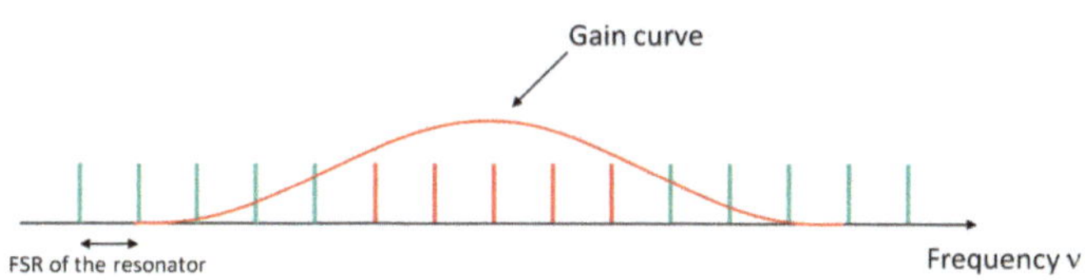

Fig. 1.9 Emission spectrum of a laser is defined by the resonator (modes) and the *gain curve* (range)

the amplifying medium then decreases when reflected off the mirrors. Thus, the number of photons gained during one roundtrip is equal to the number of photons that are lost. Following Fig. 1.10, we can then write, with P being the power of the laser just before the mirror with reflectivity R_2 and P' the expression of the power after a round trip:

$$P' = G' \cdot R_1 \cdot G \cdot R_2 \cdot \eta \cdot P$$

where η is a transmission parameter gathering all the losses on a roundtrip except those due to the cavity mirrors, and G and G′ are the effective gains in the positive (outgoing laser beam) and negative directions, respectively (Fig. 1.10). It is sometimes necessary to distinguish the effective gains G and G′ according to the direction of the wave propagation because in virtue of the gain saturation phenomenon (see Fig. 1.6) the gain G depends on the incident intracavity power, and when mirror reflectivities are very different these gains can be somewhat different.

When the laser emits continuously (steady-state is reached) $P = P'$ so the roundtrip gain $G_{rt} = GG' = 1/(R_1R_2 \cdot \eta)$.

If $G_{rt} < 1/(R_1R_2 \cdot \eta)$, the laser cannot oscillate. If

$$G_{rt} > \frac{1}{R_1R_2 \cdot \eta}$$

the power of the beam in the cavity increases at each roundtrip. The condition $G_{rt} = \frac{1}{R_1R_2 \cdot \eta}$ defines the *threshold* of the laser and gives the roundtrip gain that must be achieved in the medium to obtain oscillation. G_{rt} depends both on the pumping rate or pumping intensity I_p (for optical pumping) and on the intracavity laser intensity I, in virtue of the gain saturation mechanism exposed in Fig. 1.6. It can generally be written a relation of the type (valid for a pure 4-level system, in the CW regime):

$$G_{rt} = \frac{A.I_p}{1 + \frac{I}{I_{sat}}}$$

where A is some constant, I_p is the pump intensity (number of pump photons per s and per m^2), I the laser intensity inside the cavity, I_{sat} a saturation intensity defined and discussed in detail in Sect. 2.2.4.1. The roundtrip gain increases with the

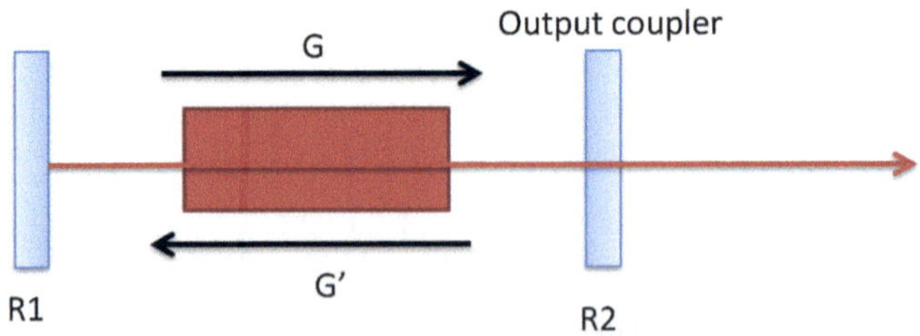

Fig. 1.10 Effect of mirrors losses and amplifying medium gain on the laser (see text).

pumping rate (here absorption saturation is neglected) and decreases with the laser intensity (gain saturation). For a given threshold gain fixed by the losses:

$$G_{threshold} = \frac{1}{R_1 R_2 \cdot \eta}$$

The intensity follows the relation:

$$I = I_{sat}\left[\frac{A.I_p}{G_{threshold}} - 1\right]$$

The laser intensity I varies linearly with I_p with a clear threshold on the pump intensity $I_{p,threshold} = G_{threshold}/A$. The existence of a clear threshold is an important distinctive behavior of laser effect.

1.6 How to Recognize Lasing?

The title of this paragraph is borrowed from I. Samuel et al. [2] who published a commentary in 2009 to properly define what should be called a laser or not, after some irrelevant claims of organic lasing in the recent literature. Indeed, recognizing proper lasing from lasing-like effects is not so simple as it appears at first sight, because it is possible to have an optical source with a narrow spectrum, good spatial coherence, even with a threshold in the light emission process, which are not however lasers.

Other processes linked to interference, Amplified Spontaneous Emission (ASE or "mirror-less lasing", that is the amplification of spontaneous emission over a waveguide for instance, leading to a relatively narrow spectrum but without mode properties) or simple waveguiding effects for example. In some specific resonator geometries, such as microcavities or waveguides, often met in organic lasers, there could be some controversy to determine if the observed emission is a laser.

Following Samuel's guidelines, a real laser is characterized by the following properties (see also Fig. 1.11):

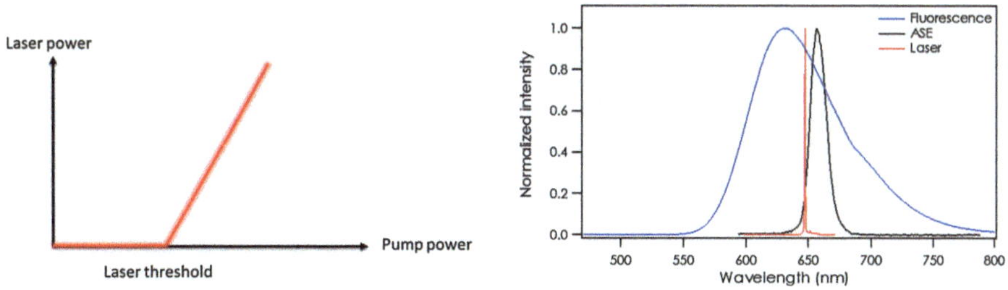

Fig. 1.11 Typical laser efficiency curve, with a clear threshold (*left*), and comparison between a laser spectrum (very narrow, in *red*) and fluorescence/ASE spectra (*right*)

1. A narrow linewidth emission;
2. A clear threshold in both output power and linewidth versus pump power;
3. The observation that the resonator geometry and the gain medium influence the light properties;

One can generally add that the generated light forms a beam, even if in some specific lasers (whispering galleries resonators for example, or nanolasers) this criterion is not fulfilled.

References

1. A.E. Siegman, *Lasers* (University Science Books, Mill Valey, 1986)
2. I.D.W. Samuel, E.B. Namdas, G.A. Turnbull, How to recognize lasing. Nat. Photonics. **3**(10), 546–549 (2009)

Chapter 2
Fundamentals of Organic Lasers

Abstract In this chapter the main characteristics and specificities of organic solid-state lasers are presented. We particularly highlight these aspects which are important for organic lasers and specific to them, and which are therefore not usually treated in classical textbooks on lasers. The objective of this chapter is to present a quite general, while not exhaustive, overview of the photophysics of organic compounds that are directly useful to understand the physics of organic lasers, as well as a theoretical framework suited to the description of these lasers in most practical situations.

In this section we first present (Sect. 2.1) the basics of pi-conjugated systems: their general properties, and why they have semiconducting properties in neat solids. We then review (Sect. 2.2) their photophysical properties: how they absorb and emit light, and why they do this very well but only in the UV–visible range. Organic lasers have a distinctive dynamical behavior compared to other gain media, in particular they are not able to work (at least up to now) in the CW regime: this peculiarity is due to the existence of triplet states, which are examined with some detail in Sect. 2.3. As organic emitters are never alone in a solid-state medium but surrounded by other emitters, of the same kind or different, in the ground or in the excited state, many intermolecular or quenching effects may arise, which strongly affect the laser performance (FRET, exciton–exciton annihilation, exciton diffusion, etc.): they are discussed in Sect. 2.4. In Sect. 2.5 we examine how these photophysical properties affect the laser operation and in particular the rate equations. Then finally in Sect. 2.6 we gather the concepts presented during all the preceding sections to deal with organic lasers dynamics, establish a condition for CW lasing and discuss the limits on the pulsewidth from practical systems.

S. Forget and S. Chénais, *Organic Solid-State Lasers*,
Springer Series in Optical Sciences 175, DOI: 10.1007/978-3-642-36705-2_2,

2.1 PI-Conjugated Molecular Systems

2.1.1 General Properties

Organic compounds consist mostly of carbon and hydrogen with a few heteroatoms such as oxygen, nitrogen or sulfur for instance. Due to the large number of binding configurations of carbon, organic materials form a huge assembly which represents several millions of reported different compounds. Not all of them, though, are suitable for lasing.

Carbon atoms have six electrons, with a ground state configuration $1s^2 2s^2 2p^2$, meaning that the *s* orbitals are fully occupied whereas two out of the three p_x, p_y and p_z orbitals are occupied by one electron each. When a carbon atom makes a bond with another atom, hybridization occurs between *s* and *p* orbitals. When the C atom connects via four single bonds to other atoms, sp^3-hybridization occurs and the four outer electrons are arranged at the corners of a tetrahedron. These bonds are qualified as σ-bonds, as they exhibit a cylindrical symmetry. Such saturated compounds are good electrical insulators. Some examples are alkanes, or saturated polymers such as polyethylene $(-CH_2-CH_2-)_n$, polystyrene $(-CH_2-CH(Ph)-)_n$ or polypropylene $(-CH_2-CH(CH_3)-)_n$.

As far as optical and electrical properties are concerned, molecules containing double or triple bonds are more complex and more interesting. Let's consider the archetypical example of ethylene ($CH_2{=}CH_2$) represented in Fig. 2.1: in this simple case, carbon atoms exhibit sp^2-hybridization. The three sp^2 hybrid orbitals are arranged in a trigonal planar geometry, pointing to the three corners of an equilateral triangle, with angles of 120° between them; the unhybridized $2p_z$ orbitals are perpendicular to this plane and overlap side-by-side to form a weaker π bond.

If the molecule presents an alternation of simple and double bonds over a planar segment, the system is said to be pi-*conjugated*. Such systems exist in many forms (small molecules, conjugated polymers, molecular crystals) that will be described in more detail in Chap. 3. The simplest examples of pi-conjugated molecules, useful for understanding the underlying physics although they are not directly useful for lasing, are benzene and butadiene (see Fig. 2.2). The overlap between the π orbitals in those structures allows for the *delocalization* of the electrons over the whole conjugated segment, meaning that the picture of the molecule as a single Lewis formula of localized single and double bonds is not valid any more. The orbitals lose their pure sp^2-hybridized character, and the molecule is instead more conveniently described by a superposition of resonant mesomeric forms. The π electrons do not belong to a single bond or atom, but rather to a group of atoms. A naïve picture of delocalization consists in thinking of a *pi*-electron moving along the conjugated chain as a "flip-flop" between single and double bonds.

The alternation of double and single bonds is a necessary condition for pi-conjugation, but not sufficient, as the bonds should also lie in the same plane to enable electron delocalization. For instance, benzene, with its 6 carbon atoms

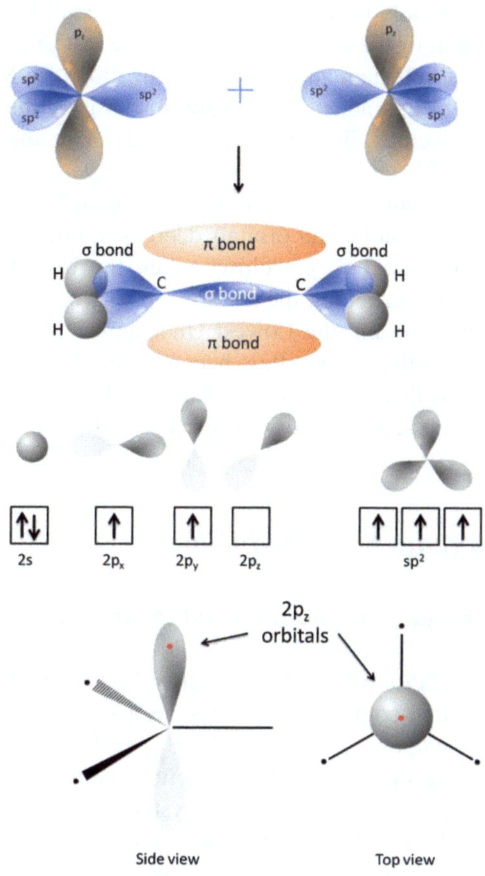

Fig. 2.1 Illustration of σ and π orbitals in ethylene

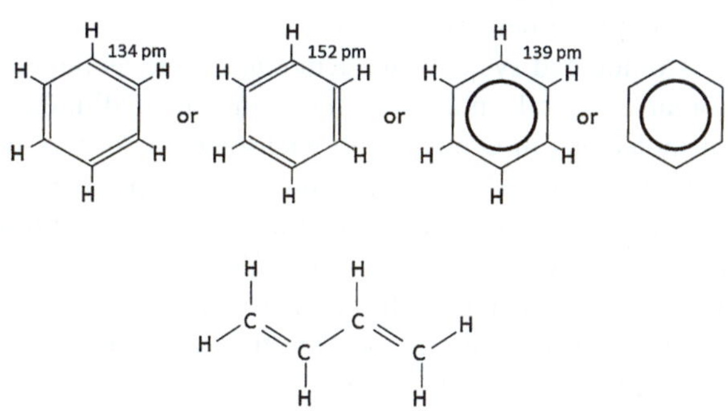

Fig. 2.2 Two simple and archetypical examples of pi-conjugated systems: benzene (*top*) and 1,3-butadiene (*bottom*). For both cases, classical Lewis formula are approximate and unsatisfactory pictures of reality, and the molecules are rather described by a superposition of different mesomeric forms

linked at the corners of a regular planar hexagon (see Fig. 2.2), is the archetype of aromatic conjugated systems. In contrast, cyclooctatetraene, which is a regular octagon with alternating simple and double bonds, adopts a "tub" nonplanar conformation and is therefore not aromatic (and only weakly conjugated). Most of practical dyes only have a portion of the molecule that is conjugated, referred to as the active unit or the chromophore unit, the rest of the molecule being for instance a saturated alkyl chain added for solubilization (see MEH-PPV for example [1]) or bulky side groups added to prevent stacking of the pi-orbitals between them in adjacent molecules, like in dendrimers [2] or in small molecules with bulky side substituents [3].

When more and more p_z orbitals are shared to form a conjugated system, the energy between molecular orbitals decreases, in a similar way as energy states of a free electron in a box get closer when the box dimensions increase (see next subsection). In particular, the energy gap between the Highest Occupied Molecular Orbital (HOMO) and the Lowest Unoccupied Molecular Orbital (LUMO) which governs the optical properties, becomes smaller. The absorption band of conjugated systems then occurs at wavelengths above 200 nm, making such compounds suitable for optical excitation in the UV–visible part of the spectrum.

2.1.2 *Organic Semiconductors*

2.1.2.1 Charge Delocalization and Polaron States

The delocalization of electrons is also the key to understand the semiconducting properties of pi-conjugated systems, in a sense which however needs some clarification. The optical gap defined by $E_{gap} = E(LUMO) - E(HOMO) - E_b$ (where E_b is the binding energy of the exciton, defined in the following) varies in the range ~1–4 eV for pi-conjugated systems, which makes them fall into the category of insulators, the "conduction band" (LUMO) of the molecule being almost empty at room temperature. Stated otherwise, there are very few free carriers available (typically less than 10^{14} cm^{-3} in organic semiconductors, where the chromophore density is around 10^{21} cm^{-3}) [4].

Conjugated systems are semiconductors as far as we consider molecular solids (organic crystals, polymers, neat films of small molecules…) made of an assembly of molecules in contact with each other, either in an ordered or disordered medium. The term "contact" just means here that there is a possibility for a charge to travel from a molecule to another, a loose definition which excludes dye-doped nonconductive polymers or dye-doped glasses, and obviously dyes in liquid solutions or gas phase. In order to understand the semiconducting properties, we have to consider that among the molecular assembly we have a *doped* molecule, that is, with an additional or a missing electron on it. Indeed electrical conduction makes sense only whenever one of the bands is partially empty; here it translates in saying that the HOMO or the LUMO level contains only one electron instead of two. From a chemist's point of view, these states are oxidized or reduced forms of the molecule, namely charged radical ions. From the physicist's point of view, the same objects are often referred to as "holes" (when one electron is missing in the HOMO) or "electrons" (when one extra electron is added in the LUMO); however the adequate word to describe these states is a *polaron*. First of all, as organic pi-conjugated molecules are rather big molecules (with molecular weights over a few hundreds for the simplest dyes) they have many vibrational modes and may be thought of as a complex system of springs coupled together in a very classical mechanics point of view. Adding or removing a charge on such a molecule may then not happen without a significant vibrational relaxation whose aim is to seek the new free energy minimum that will take into account the presence of the extra charge. Furthermore the nearby neutral molecules will react at the presence of a charge by a slight displacement of their inner charge clouds, creating a macroscopic electrical polarization around the doped molecule, which in turn will lead to a further stabilization of the latter. A *polaron* is the quasi-particle corresponding to the charged molecule and its accompanying polarization field; the electron polaron state has a lower energy than the genuine LUMO state (basically it means that more energy is required to pull the electron out of the molecule than it would be for an isolated molecule in gas phase), and symmetrically the hole polaron state lies higher in energy than the original HOMO (see Fig. 2.3), by an amount called

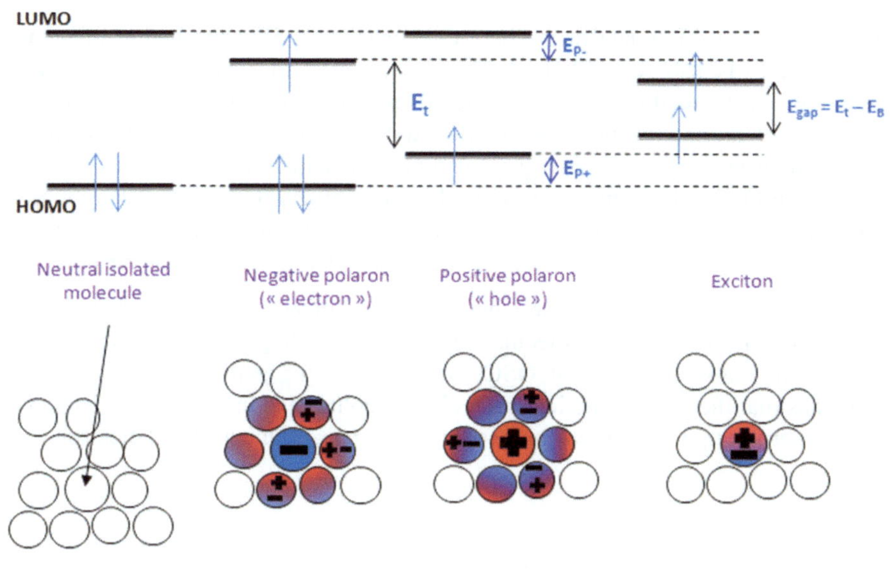

Fig. 2.3 Illustration of relevant energy levels in organic semiconductors. The HOMO and LUMO states pertain to isolated molecules, in gas phase or in a nonpolar solvent/nonpolar solid matrix. In neat solid materials where the molecules form organic semiconductors (OSC), the neighboring polarizable molecules create a charge cloud around the charge which modifies the energy levels of the molecular ion by an amount E_{P+} (for holes) and E_{P-} (for electrons): the whole entity is called a polaron. In an OSC the effective "transport gap" between electrons and holes is defined as $E_t = E(LUMO) - E(HOMO) - E_{P+} - E_{P-}$ [99]. The optical gap, usually measured at the onset of optical absorption, includes an exciton binding energy term E_B, because it requires some extra energy to separate a pair of coulombically-bound charges onto different molecules ($E_{gap} = E_t - E_B$)

the polaron binding energy, whose typical order of magnitude is a few tens of eV (0.137 eV calculated in anthracene [5] or 0.29 eV in TPD [6]).

This charge added on the molecule is also delocalized over the conjugation segment and can be thought of as a mobile charge. A mobile charge on a molecular moiety does not still make the medium a conductor, because in order to carry electrical current, charges must be able not only to travel within one molecule, but also to pass from a molecule to another and join electrodes. In a disordered medium (that is, most of practical polymers or small-molecular films used for organic optoelectronics applications) this is usually done by "hopping", basically through tunneling, from a conjugated site to another, or in a more classical "bandlike" conduction picture in the case of organic crystals. The hopping mechanism is highly inefficient and clearly the limiting factor which explains the very low carrier mobilities in organic semiconductors compared to inorganic semiconducting crystals. Details about charge transport in OSCs can be found in Refs. [7–12].

2.1.2.2 Excitons

In addition of a difference in the nature of charge carriers, the nature of neutral excited states is also very different. These states are called *excitons* because they have the ability to diffuse from a site to another and hence comply with the definition of an exciton being a mobile excitation quasi-particle. The main difference between excitons in inorganic and organic semiconductors arises from the fact that in OSCs electrons are only weakly delocalized over a small length scale even in the case of molecular crystals or conjugated polymers. As a result they have two important peculiarities as compared to their inorganic counterparts.

1. The first one is a considerable binding energy, linked to the strong localization of excitons in a single molecular site (and accordingly to a small dielectric material constant $\varepsilon_r \sim 3$ in OSCs versus ~ 13 for gallium arsenide, which means that there is no efficient screening of columbian interactions), defined as the energy required to break an exciton into a pair of uncorrelated electrons and holes:

$$E_b = \frac{e^2}{4\pi\epsilon_0\epsilon_r d_{e-h}}$$

where $d_{e\text{-}h}$ is the separation distance of the charge carriers. The typical values are 0.5–1.5 eV for organic crystals [13] and usually less (0.2–0.5 eV) for conjugated polymers [14]. In OSCs where electrons and holes can hop from a site to another, this strong binding energy means that excitons are particularly stable against charge separation. The situation in terms of energy levels is summarized in Fig. 2.3.

2. The second difference is the existence of well-defined spin states (singlet and triplet) for excitons which makes them no different from isolated molecules in this respect. The triplet excited states have usually a lower energy level than the singlet excited state by an amount called the exchange energy ΔE_{ST}, which will be discussed in more details in Sect. 2.3.2.

2.1.2.3 Is There a Difference Between a "Dye" and an "Organic Semiconductor"?

Before discussing this topic, we may try to define properly these two terms. The first organic lasers were developed in the mid 60's by Sorokin and Lankard [15] in liquid solutions and called "dye lasers". The term "dye", inherited from the chemical and textile industry, was at that time used to designate any kind of luminescent π-conjugated system [16], but used in practice only for highly-luminescent small molecules in organic solvents. Organic electronics later emerged together with the notion of organic semiconductor (OSC), which embraces a much broader family of materials, in particular luminescent conjugated polymers. The term of OSC actually refers to any kind of conjugated system where charge carriers are able to travel from a *pi*-system to another, even though this should happen with a low mobility: this means that the *pi* orbitals of two neighbouring sites have to overlap in some way. The term OSC hence describes a large class of materials but does *not* comprise conjugated systems in solution, either liquid or solid, where emitters are disseminated in a nonconductive medium, i.e. a liquid solvent or a polymer such as polymethyl methacrylate (PMMA) for instance. Lasers based on the latter are still referred to as "(solid-state) dye lasers" in the literature [17]. Everyday use reveals that it is uncommon to hear of a conjugated polymer or an organic crystal as a "dye". Conversely fluorescent small molecules are equally referred to as "dyes" or OSCs, depending whether they are used disseminated within a host matrix, or in pure *neat films*, respectively. In general, it is observed that small fluorescent molecules that are highly luminescent in dilute form become non-emissive in neat films because of the concentration quenching phenomenon (see Sect. 2.4). However, there are numerous examples of molecules where this is not the case, for example many emitters used in Organic Light-Emitting Diodes (such as N'-diphenyl-N,N'-bis(1-naphthyl)(1,1'-biphenyl)-4,4'diamine known as NPB, or Tris(8-hydroxyquinolinato)aluminium known as Alq_3, both used in the first generation of OLED devices). In these cases, as far as quenching does not totally hinder light emission, the distinction between dye and OSC becomes rather irrelevant. A more detailed discussion on this terminology can be found in [18] and is discussed from the point of view of materials classification in Sect. 3.1.2.

2.2 Photophysical Properties of Pi-Conjugated Systems

2.2.1 Absorption of Light by Pi-Conjugated Systems

2.2.1.1 Energy States Involved in Optical Transitions

Optical transitions associated with *single* bonds are of the type σ–σ^* and correspond to absorption bands at wavelengths below 160 nm: the related photon energy (>7 eV) is higher than the dissociation energy of most other chemical bonds, meaning that photochemical decomposition is likely to occur, making such compounds not suitable for laser action. Most of laser materials will consequently be based on π–π^* transitions. In some molecules, non-bonding orbitals are present, denoted as n orbitals. A typical example is carbonyl compounds where there are two electrons in the non-bonding 2p orbitals of the oxygen atom. Absorption of radiation can occur in a $n-\pi^*$ transition, i.e. from a "localized" bond to a delocalized bond.

For a convenient description of photophysics, energy states are labeled depending on an important parameter, the spin multiplicity. As described in more detail in Sect. 2.3, electrons have spin angular momentum, described by a spin quantum number $s = \pm ½$. The sum of all spins of all electrons involved in a system defines the spin S of the system, and spin multiplicity is defined as $2S + 1$. When $S = 0$ the state is said to be a singlet state; when $S = 1$, the multiplicity is 3 and the state is said to be a triplet state.

Excited states in organic materials are created upon absorption of a photon, that is promoting one of the two electrons of the ground state to one empty higher-lying orbital. As inner, filled, energy states are not involved in optical processes, the ground state of the system is usually described by a single filled orbital (the HOMO) containing two electrons, whose spins are necessarily antiparallel in order to comply with the Pauli exclusion principle. As a consequence, the ground state of an organic system is almost always a singlet state, and is denoted as S_0. A notable exception is molecular oxygen, whose ground state is a triplet state, since in this quite exceptional case, the ground state results from two unpaired electrons in degenerate orbitals.

Excited states are described by two electrons: one residing in the HOMO level, the other one in another antibonding molecular orbital, which may be the LUMO or any another higher-lying antibonding orbital. Because the two electrons are not any more in the same electronic level, Fermi exclusion principle does not hold any more and all spin combinations are possible. The singlet excited states are labeled S_1, S_2, etc. and the triplet excited states are referred to as T_1, T_2, etc.

Not all optical transitions are possible, they are restricted by optical selection rules [19]. The most important in organic photonics is the *spin selection rule*: for a pure electric dipolar transition, the electric field of the incoming radiation is not able to switch the spin of an electron, so that absorption from the ground singlet state S_0 can only lead to excited singlet states S_1, …, S_n. Creating triplets is not though totally forbidden, and can be enhanced through spin–orbit coupling. Refer to Sect. 2.3 for more details.

Other selection rules are the orbital symmetry selection rule (the two orbitals describing the ground and the excited state must resemble each other), and the parity rule (only transitions between states with different parities are allowed). More details can be found for example in [20].

2.2.1.2 The Energy Band Gap in Pi-Conjugated Systems

Light absorption by conjugated systems can be understood qualitatively with a simplified quantum–mechanical model, such as the free-electron gas model [16]. This simple analogy is illustrated in Fig. 2.4 with the simplest possible linear conjugated system, S-*trans* 1,3 butadiene (C_4H_6). The molecule is planar; carbon atoms connect with other C atoms and with the six hydrogen atoms through three σ-bonds each, meaning that there is a total of 18 sigma Molecular Orbitals (Nine MO of type σ occupied by two electrons, and nine unoccupied σ^* orbitals) for which the highest probability to find the electrons occurs within the molecular plane containing the nuclei (single bonds).

In contrast, the four remaining π-electrons occupy orbitals with one node in the plane of the molecule, and form a charge cloud above and below this plane. It is informative to interpret the two lobes (and the related zero probability to find a pi-electron in the molecular plane) as the result of a destructive quantum interference, evoking the $n_z = 2$ quantized mode of a quantum box in the z-direction (see Fig. 2.4): they represent a mode with one node at the center, having opposite signs of the wavefunction on the two sides. The box modes represented in Fig. 2.4 are strictly valid for a constant potential inside the box which is obviously not the case especially in the plane of the molecule, but the physical contents remain. Thus, with the simple picture of a rectangular box with infinite walls, the four pi-orbitals plotted in Fig. 2.4 resemble the four first quantized modes $n_x = 1, 2, 3, 4$ with the fixed quantum numbers $n_z = 2$ and $n_y = 1$. A linear conjugated system appears as a nanowire with discrete energy levels due to confinement. If we simplify a little bit further and consider a linear molecule as a 1D system of length L, then the energy $E(n)$ of the nth eigenstate is given by

$$E(n) = \frac{h^2 n^2}{8 m_e L^2}$$

where n is the quantum number (equal to the number of antinodes along the chain), m_e the mass of the electron, h the Planck's constant. When some light is incident upon a conjugated molecule, an optical transition occurs between a filled orbital and an empty one. The lowest photon energy enabling such a transition defines the *optical gap*, which is somewhat lower than the *HOMO–LUMO* gap. The latter is defined as the energy difference between the **H**ighest **O**ccupied **M**olecular **O**rbital

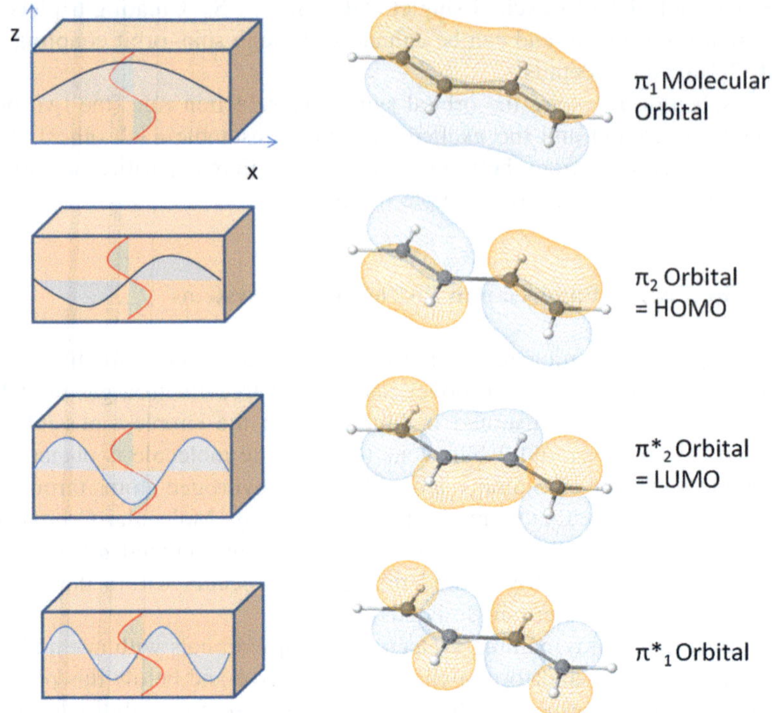

Fig. 2.4 Analogy between the electronic levels of a free electron in a rectangular quantum box with infinite walls (*left*), and the pi molecular orbitals of S-*trans* 1,3-butadiene, calculated with Orbimol® [21]. The pi molecular orbitals correspond to a zero probability to find the electrons within the molecular plane; in the quantum box model, although this corresponds to the nonrealistic case of a constant potential inside the box, π-orbitals can be ascribed to $n_z = 2$ modes (one node in the plane cutting the box in two parts in the z-direction), the σ orbitals corresponding to $n_z = 1$. Since butadiene contains 4 pi-electrons, they occupy by pairs the two first π_1 and π_2 orbitals, which are associated to $n_x = 1$ and $n_x = 2$, respectively, where n_x is the number of antinodes in the longest direction of the quantum box. The LUMO and LUMO + 1 levels, empty in the ground state, correspond to the $n_x = 3$ and $n_x = 4$ modes

(HOMO) and the **L**owest **U**noccupied **M**olecular **O**rbital (LUMO); the difference between the HOMO–LUMO gap and the optical gap arises from the attractive interaction existing between a hole (absence of electron in HOMO) and an electron (supplementary electron in LUMO) when they coexist in a single molecule to form the excited state; this is actually the exciton binding energy evoked above, added to the exchange energy specifically for triplet states (see Sect. 3). However we neglect for the moment this difference for the sake of simplicity. Generally the number of *pi*-electrons N is an even number (this is true for stable molecules, notable exceptions are polarons, or charge carriers, in OSCs) which means that the

N/2 first π orbitals are full, up to the HOMO orbital included, and the N/2 upper π^*-orbitals are empty starting from the LUMO. Then the energy gap writes:

$$E_{gap} = E(LUMO) - E(HOMO) = E\left(n = \frac{N}{2} + 1\right) - E\left(n = \frac{N}{2}\right) = \frac{h^2(N+1)}{8m_eL^2}$$

This minimal energy gives the longest wavelength that can be absorbed, through the relation:

$$\lambda_{max} = \frac{hc}{E_{gap}} = \frac{8m_ecL^2}{h(N+1)}$$

This indicates that within a good approximation the position of the absorption band is determined by the chain length *L* and the number of π-electrons *N* only. In the example of butadiene, $E_{gap} = 9.73$ eV (calculated from HOMO–LUMO levels [21]), which yields with N = 4 an expected chain length L = 4.4 Å. Each of the three bonds in S-trans-butadiene is 1.35 Å long, meaning that there is a fair agreement between the simple theory and the experimental values provided that the chain length L extends beyond the terminal atoms by only 0.18 Å on both sides. In general, there is a correct match for linear simple dyes (e.g. cyanines [16]) but in most practical cases unfortunately the relationship between chemical structure and absorption spectrum is not so simple.

For benzene for instance, in which the six *pi*-electrons are fully delocalized above and below the six carbon atoms, naïve pictures do not work so well: attempting to retrieve the energy gap from the circumference of the ring (let's say, in a similar fashion as we do in the Bohr model to retrieve the electronic levels of the hydrogen atom) leads to large errors because the repulsion force between the six electrons is strong and plays a prominent role [16].

From this simple picture however we can approach how it is possible to tailor the spectral properties of pi-conjugated systems. It appears relatively easy to modify the absorption spectrum (and in a closely related way, the emission spectrum) by varying the chemical structure. This sets a huge (and very useful in practice) difference with inorganic semiconductors where the emitted wavelengths are strongly limited by the available materials and by lattice-matching restrictions.

There have been many efforts made over the past decades, in molecular engineering, to achieve such a "bandgap engineering" of *pi*-conjugated systems [22, 23]. Playing on the conjugated length is the most straightforward way, as illustrated in the polyacene series, where the adjunction of phenyl rings from benzene to pentacene leads to a redshift of absorption spectrum [24]. It is not the only way to tune the bandgap: another efficient strategy (illustrated in Fig. 2.5) consists in incorporating Electron Withdrawing Groups (EWG) like keto, cyano or dicyano groups which tend to lower the LUMO energy level, or conversely to attach Electron Donating Groups (EDG) such as alkyl, alkoxy or alkylsulfanyl

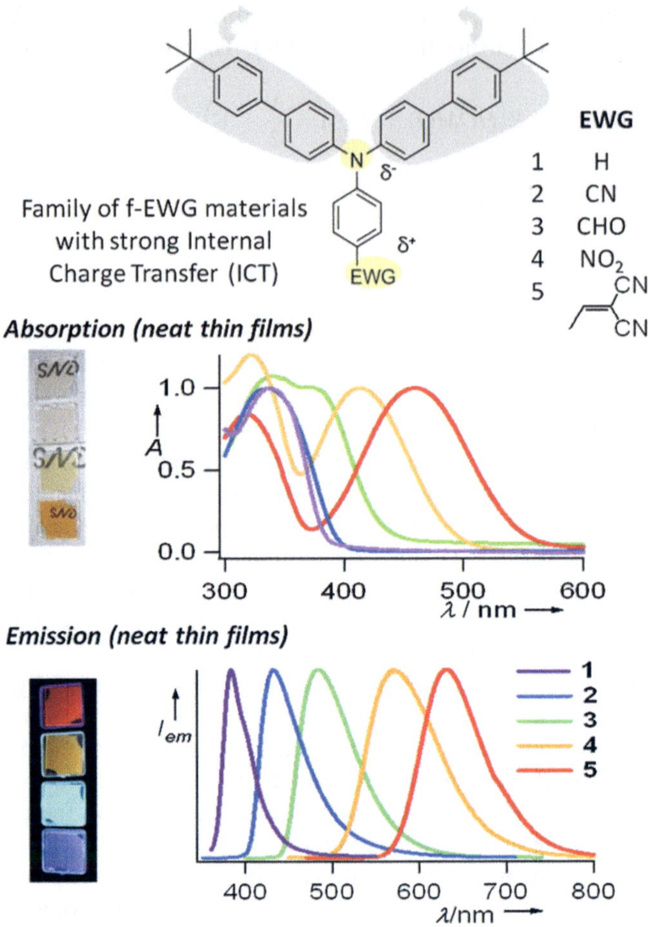

Fig. 2.5 Illustration of bandgap engineering and chemical tuning by substitution of different Electron-Withdrawing Groups (EWG) on a triphenylamino moiety. The EWG are labeled from 1 to 5 from the lowest electron-accepting to the highest electron-accepting character. The more the EWG attracts electrons, the lower the bandgap and the more the absorption and emission are redshifted. Absorption and fluorescence spectra of 100-nm-thick evaporated neat films are shown. The strong light emission by neat solid films of dye molecules is a rather uncommon feature, because in general concentration quenching prevents emission. Here it can be explained by several factors: low pi-stacking thanks to steric hindrance of tert-butyl terminal groups and a twist of the triphenylamino core, as well as a strong Stokes shift preventing intermolecular dipole–dipole or radiative energy exchange. Courtesy of E. Ishow. Reprinted with permission from [3]. Copyright (2008) American Chemical Society

groups which tend to elevate the HOMO energy level [23]. In both cases the tendency is to lower the bandgap. The spectral limits to the UV and to the IR of bandgap engineering are discussed in Sect. 3.2.

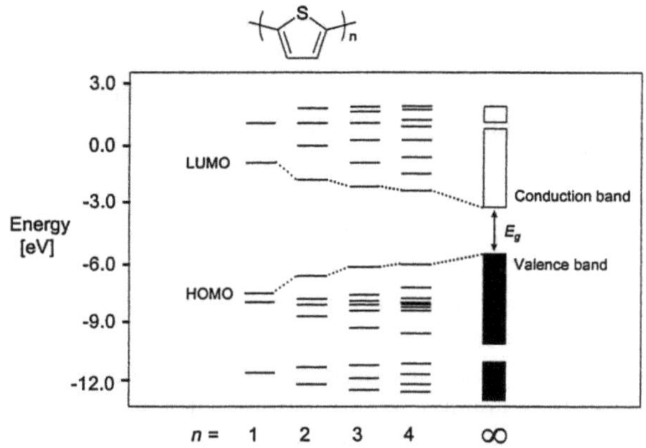

Fig. 2.6 Calculated (frontier) energy levels of oligothiophenes with n = 1–4 and of polythiophene, where E_g = band gap. Courtesy of E.W. Meijer. Reprinted from [100], Copyright (2001), with permission from Elsevier

2.2.1.3 Absorption Spectra

Origin of Broad Spectra

A striking peculiarity of the absorption spectra of organic systems, as opposed to atomic/ionic spectra, is the width of the bands, usually covering several tens up to a hundred nanometers. The large number of electronic individual *pi*-orbitals shared in a *pi*-conjugated system tends to form "bands", as sketched in Fig. 2.6 for the typical example of oligothiophenes.

However this image of band formation does not explain completely why absorption spectra are *continuously* broad, and not formed of multiple sharp peaks that we could resolve with a high-resolution spectrometer, for instance, especially for small molecules where the number of shared *pi* orbitals is not so large.

It becomes clear when one considers that even a small dye molecule (such as DCM, represented in Fig. 2.6) contains several tens of atoms, with at least a hundred or more associated vibration modes of the molecular skeleton. These vibrations of the nuclei densely cover a spectrum between a few cm^{-1} to $\sim 3000\ cm^{-1}$. Indeed, it is useful to keep in mind that when a simple classical oscillator (e.g. a spring) with a given natural resonance frequency is coupled to an identical resonator (through a loose spring for instance), then the two coupled resonators have two eigenmodes whose eigenfrequencies are separated by an amount representing the strength of the coupling [25]; here the numerous possible vibrations inside a molecule couple each other and give birth to a dense ladder of closely-spaced vibrational states. These vibrational states are coupled to electronic states, so that each electronic state is in fact a set of closely-spaced *vibronic* levels, since basically the energy of an electron in a molecular orbital depends on its position with respect to the nuclei, which are all subject to vibrational motion. For completeness, let's add that each vibrational level is broadened by the existence of many close rotational levels (the term ro-vibrational level is hence used to described the coupled states), and that a large molecule has many geometrical conformations, each of which being associated to a slightly different position of the resulting vibronic levels. As a result the spectra of dye molecules and organic semiconductors are very broad (~ 100 nm) and continuous, which makes them very valuable for tunable lasers.

Homogeneous Broadening

An important aspect of this broadening related to laser applications its nature. In organic systems *the broadening is homogeneous*, meaning that all molecules taken individually have the same spectrum than the ensemble, so that every molecule can contribute equally to the building of a given oscillating laser mode. Stated differently, the broad bands observed in *pi* systems do not result from the superposition of many narrow lines coming from different emitters, like it is the case for instance in a Doppler-broadened gas medium, where contributions to the gain at a given wavelength is ascribed to a class of molecules having a given velocity in the gas. However there may exist some sources of inhomogeneous broadening, resulting from the conformation distribution and from variations of the local environment seen by the chromophore.

Absorption Cross Sections

Absorption is usually quantified by the absorption coefficient α (in cm^{-1}) or preferably by the *absorption cross section* $\sigma_{abs} = \alpha/N$, a figure which is however meaningful only if the density of absorbing units N (in cm^{-3}) is well defined or can be estimated. The absorption cross section relates to a single molecule, it has the dimensions of a surface (usually expressed in cm^2) because it represents the effective area of the molecule as if the latter were a totally-absorbing planar element of area σ_{abs} set in the path of the beam. The (unsaturated) transmission $T(\lambda)$ of a homogeneously absorbing sample of length L is simply given by:

$$T(\lambda) = exp[-\alpha(\lambda)L] = exp[-\sigma_{abs}(\lambda)NL] = 10^{-A(\lambda)}$$

where A is the absorbance (usually defined in base 10, with $A = \varepsilon(\lambda)\ cL = \frac{\alpha(\lambda)L}{\ln(10)}$, where $\varepsilon(\lambda)$ is the molar absorptivity in $M^{-1}\ cm^{-1}$ or liters/(mol cm), and c the concentration in M). Note that the validity of the Beer-Lambert law is restricted to linear absorption in absence of stimulated emission.

Although this is not the most convenient unit for a laser physicist, the molar absorptivity ε is often found as a measure of absorption of molecular systems in the literature, and is simply connected to the absorption cross section by:

$$\sigma_{abs}\left[cm^2\right] = \frac{1000\ln(10)}{\mathcal{N}_A}\varepsilon\left[l\,mol^{-1}\,cm^{-1}\right] = 0.385.10^{-20}\varepsilon\left[l\,mol^{-1}\,cm^{-1}\right]$$

where $\mathcal{N}_A$ is the Avogadro number.

Experimental determination of absorption coefficients $\alpha(\lambda)$ is straightforward. It is evaluated from the spectral transmission measurement $T(\lambda)$ performed with a spectrophotometer operating at low intensity input light levels to work in the linear absorption regime. Absorption cross sections can then be derived as far as the chromophore density N is known. The latter parameter is simple to estimate in the case of dye-doped polymer films: the exact amount of dye in the film is deducible from the mass doping ratio x and the density of the film ρ. The latter can be reasonably approximated to the density of the undoped polymer film, as far as the doping ratio does not exceed a few percents. The dye density is then equal to:

$$N = \frac{x\rho\mathcal{N}_A}{M}$$

where M is the molar mass of the dye. From this relationship we can derive the useful following relation, which enables computing the absorption cross section of a dye molecule directly from the absorbance measurement of a polymer film or rod of density ρ and length L, doped at a mass ratio x with a dye of molar mass M:

$$\sigma_{abs}\left[cm^2\right] = \frac{M[g\,mol^{-1}]A(\lambda)\ln(10)}{x\mathcal{N}_A\rho[g\,cm^{-3}]L[cm]}$$

In organics the absorption cross sections are large, typically on the order of $\sigma_{\text{abs}} \sim 10^{-15}$ cm^2. They are comparable or might be slightly lower than absorption cross sections for interband transitions in inorganic semiconductor quantum dots (typically $>10^{-14}$ cm^2 [26]), but they are considerably larger than cross sections of rare-earth ions in crystalline laser media, which are usually based on metastable transitions that are forbidden in nature, and happen to be allowed in a solid-state environment in virtue of the coupling with the solid matrix; thus, in Neodymium, Ytterbium or Erbium based lasers, cross sections are only[1] typically $\sim 10^{-19}$ cm^2.

In conjugated polymers or organic crystals, the density of active units or the number of chromophores per chain is not precisely known; in these cases the absorption coefficient α might be more relevant to quantify absorption than the absorption cross section. α has typical values of 10^4–10^5 cm^{-1} at the peak absorption wavelength in organic solids, which is comparable to values measured in bulk inorganic semiconductors for interband transitions near the band gap [27].

In the case of neat films (x = 1) of evaporated small molecules, N is of the order of $N \approx \rho \times N_A/M = 1.3 \times 6.10^{23}/460 \sim 10^{21}$ cm^{-3} where ρ is the film density (numerical values here are those of Alq_3 [28]), which means that the

[1] Although this low absorption cross section is also associated to a low emission cross section (and hence a low gain), this does not mean that crystalline gain media doped with rare-earth ions would be "worse" than organics or semiconductors to make a good laser material, because the metastability of the excited state also means that Nd, Yb or Er ions have a very long lifetime ($\sim$ms), enabling a good energy storage capability. For this reason population inversion is easy to reach in rare-earth ions, even in three-level (Er) or quasi-2 level (Yb) configurations.

incident light will be absorbed over a typical length scale of $L_{abs} = \alpha^{-1} = (\sigma_{abs}N)^{-1} \sim 100\,\text{nm}$. This is also the typical order of magnitude for a film of conjugated polymer like PFO or PPV. This value is noticeably lower than the optical wavelength, which will have some impact on the laser resonator design (see Chap. 4).

For theoretical work on lasers based on conjugated polymers, it may be however useful to know the absorption cross section. In this case, it is possible to derive it indirectly, by measuring first the fluorescence quantum efficiency and the radiative lifetime, and then obtaining the absorption cross section using the Strickler–Berg formula exposed later in Sect. 2.4 [29].

The Transition Dipole Moment

The absorption cross section spectrum is not the single figure needed to properly quantify absorption in organic *pi*-conjugated systems; indeed absorption is fully described by a vectorial quantity, the transition dipole moment.

The absorption transition dipole from the ground state g to an excited state e is defined by [30]:

$$\overrightarrow{d_a}(g \rightarrow e) = \left\langle \psi_g \middle| \tilde{\vec{d}}_a \middle| \psi_e \right\rangle$$

where ψ_f and ψ_e are the stationary wavefunctions of the involved states, and $\tilde{\vec{d}}_a$ is a vector operator being the sum of the position vectors of all charged particles weighted with their charge. According to the Franck–Condon principle, absorption is a fast process and the positions of the nuclei are thus assumed to be fixed, so that only the electronic part of the wavefunctions is affected during the absorption of a photon. In the most simple and archetypal case of a dye planar aromatic molecule with a $S_0 \rightarrow S_1$ transition having a π–π^* character, both ψ_g and ψ_e orbitals are symmetrical with respect to the plane of the molecule, so that the integral defining $\overrightarrow{d_a}$ along the direction perpendicular to this plane is zero, since the integrated function is odd. Hence $\overrightarrow{d_a}(S_0 \rightarrow S_1)$ lies in the plane of the molecule. If the dye is pumped in its S_2 state, the excited wavefunction involves in general a π^* orbital as well, and $\overrightarrow{d_a}(S_0 \rightarrow S_2)$ remains in the plane of the molecule. However the profile of the π^* orbitals in the plane are different, and thus their corresponding transition dipoles are not oriented similarly.

In the case of a "rod-like" or linear molecule such as DCM (see Fig. 2.7) the absorption transition moment will be along the molecule axis [31]. In polyacenes (naphthalene, anthracene…) the $S_0 \rightarrow S_2$ moment is perpendicular to the $S_0 \rightarrow S_1$ moment, the latter being along the axis connecting the phenyl rings [32]. This is also the case of fluorescein-based molecules represented in Fig. 2.8.

In optical experiments, this feature means that a molecule will respond differently according to the polarization state of the incoming radiation. In solid films

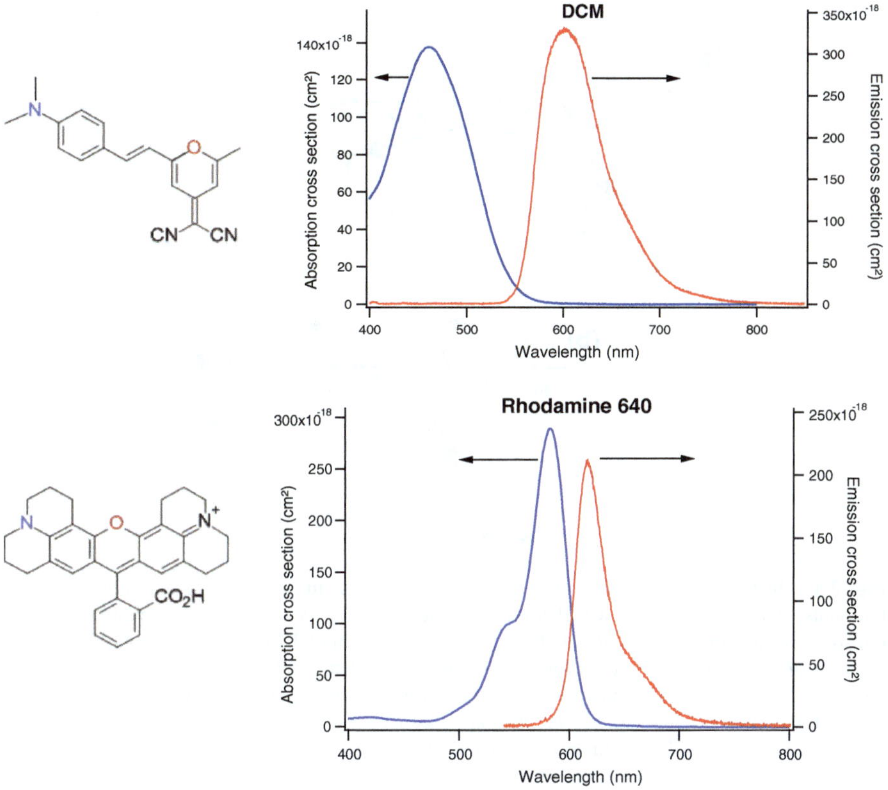

Fig. 2.7 Absorption and emission cross-section spectra of DCM and Rhodamine 640 in PMMA

where molecules are not randomly oriented, this causes the absorption to be polarization-dependant [33].

2.2.2 *Emission of Light by Pi-Conjugated Systems*

2.2.2.1 Emission Spectra

Let's now examine the fate of a molecule after absorbing a photon of energy hν. Absorption occurs between the lowest-lying vibrational state of the S_0 electronic state and a vibronic subband of the S_1 or S_2 excited singlet states, as selection rules make the singlet-to-triplet transitions forbidden, and therefore associated to very low absorption cross sections compared to singlet-to-singlet transitions.

Like absorption spectra, emission spectra are broad and have a typical width at half maximum of several tens of nm. This broad fluorescence leads to a possible tuning of the emission wavelength, thanks for instance to a dispersive element

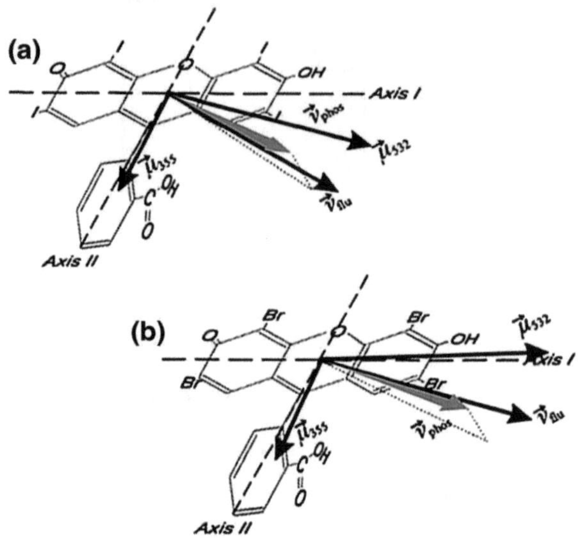

Fig. 2.8 Illustration of the difference which may exist in the orientation of dipole transition moments, depending on the wavelength of excitation. The two examples correspond here to phosphorescent small molecules with a fluorescein core and I or Br substituents. The dipole moment of absorption is featured for an absorption wavelength of 532 nm (S0 → S1 transition) and at 355 nm (S0 → S2 transition), and they appear to be almost orthogonal. The transition dipole for emission (fluorescence or phosphorescence) is aligned with none of the absorption moments. Courtesy of P. Lettinga. Reproduced from [101]

added inside the laser cavity [34], and also makes them capable of ultrashort pulse generation [35].

A key point for organic laser physics resides in the existence of a large *Stokes shift* in organic materials, i.e. a redshift of the emission compared to the absorption.

In virtue of the Franck–Condon principle [36][2] the electrons in a molecule are supposed to move much faster than the nuclei, because of their much lower mass. Hence, when a molecule absorbs a photon, with a timescale of 10^{-15} s, the entire molecular structure takes some time ($\sim 10^{-12}$ s) to relax and thermalize to the minimum vibronic level of a S_1 state whose minimum lies at a higher normalized nucleus coordinate r. A physically-intuitive picture of such a relaxation can be drawn with the simple example of the H_2 molecule, explained in many elementary

[2] "Classically, the Franck–Condon principle is the approximation that an electronic transition is most likely to occur without changes in the positions of the nuclei in the molecular entity and its environment. The resulting state is called a Franck–Condon state, and the transition involved, a vertical transition. The quantum mechanical formulation of this principle is that the intensity of a vibronic transition is proportional to the square of the overlap integral between the vibrational wavefunctions of the two states that are involved in the transition"—IUPAC Compendium of Chemical Terminology, 2nd Edition (1997).

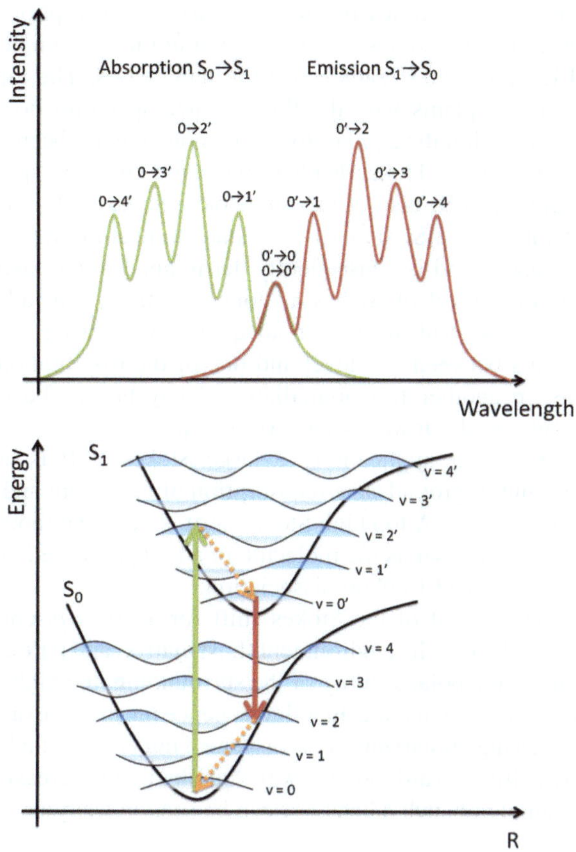

Fig. 2.9 Illustration of the Stokes shift and mirror symmetry of absorption and emission spectra. The molecular potentials are sketched as a function of R, a spatial coordinate linked to the position of the nuclei. Vibronic eigenstates are denoted by *v* and *v*'

textbooks [36, 37]. In an energy E versus r (nuclear coordinate) diagram, optical transitions then occur vertically, as sketched in Fig. 2.9, where the typical Morse-shaped molecular potentials reminds us that at high vibrational levels the stretching of bonds can lead to molecular dissociation. As explained above spectra recorded at room temperature do not allow resolving the vibronic levels in typical systems, but at cryogenic temperatures, one may observe sharp peaks and check for the "mirror-image" symmetry property between absorption and emission spectra [16].

Typical examples of absorption/emission spectra are given at room temperature in thin solid films of PMMA in Fig. 2.7 for two well-known red emitting organic dyes, namely Rhodamine 640 and DCM.

2.2.2.2 A 4-Level Sytem (At First Sight)

For lasers, the direct consequence of the existence of a Stokes shift is that *organic lasers are at first approximation 4-level systems*. The presence of higher-lying singlet states (allowing for excited-state absorption), triplet states, and in some cases polaron states makes the medium more complex but the approximation of a 4-level system is satisfactory in many cases. The rapidity of the intraband decay ($\sim$ps) explains why the fluorescence spectrum does not depend on which sub-level (within the S_1 manifold) the molecule has been excited. Excitation in a higher level S_n ($n > 1$) also leads to the same emission spectrum, as internal conversion from S_n to S_1 also occurs on a picosecond scale through nonradiative processes. Pumping to a S_n state ($n > 1$) leads to heat dissipation through these nonradiative channels, and is consequently detrimental to the laser efficiency. The lower-lying vibronic level of S_1 has a "long" lifetime, typically a few nanoseconds, and is from this point of view an adequate upper lasing level. The lasing transition then occurs between this level and one of the ro-vibrational levels of the S_0 manifold, before another fast nonradiative decay brings the molecule back to the ground state, i.e. the lower-lying level of S_0.

For laser applications, a large Stokes shift is desirable as it minimizes the amount of ground state reabsorption of laser light and makes is closer to a true four level system. A too high Stokes shift is not ideal though, as it enhances the thermal loading and sends the molecule in a highly-excited vibrational state which is more likely to lead to photodissociation.

The extent of the Stokes shift for a given *pi*-conjugated compound is highly dependent on the environment (host matrix) and particularly on its polarizability: the higher the polarizability of the surrounding, the higher the redshift [32]. This well-known effect in molecular fluorescence means for example that when a dye molecule with a high polarizability is dispersed into a nonpolar host matrix, the emission curve will shift towards longer wavelengths upon increasing the dye concentration. An example of such a behavior can be seen in triarylamines in Alq_3 [38].

2.2.3 Jablonski Diagrams

Photophysical events within a chromophore moiety (flows between energy states for a given dye molecule or chromophore unit in a polymer) can be sketched with the help of molecular potentials as presented in Fig. 2.10. Plotting the vibrational ladders inside the typical Morse potentials would be helpful for the understanding of some photodissociation effects occurring when molecules are excited to highly-excited vibrational states for instance. However, experimental access to the exact shape of these potentials is impossible in virtue of the numerous geometrical conformations of one molecule in a solid, each conformation being associated to one molecular potential. A simpler and much more widespread picture consists in drawing the energy levels as simple horizontal lines, in a Jablonski diagram (see Fig. 2.11).

Two groups of electronic states exist: singlet states (S_0, S_1,..., S_n) which have a total spin of zero, and triplet states (T_1, T_2,..., T_n) which have a total spin of one.

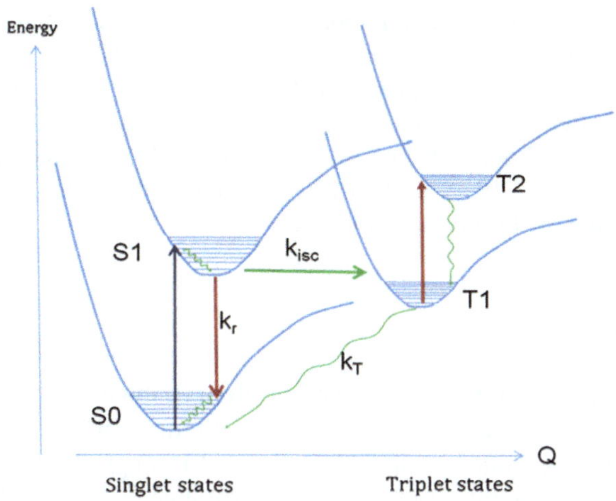

Fig. 2.10 Energy level diagram relevant for organic dyes versus a normalized nucleus coordinate Q. The shift in Q between singlet and triplet states is pictured for clarity only. In organic semiconductors more complex intermolecular (or intrachain) interactions occur. Lasing occurs on the singlet $S_1 \rightarrow S_0$ transition, where stimulated emission (SE) can efficiently bypass the radiative spontaneous emission occurring with a rate k_r. Intersystem crossing (ISC) causes triplet states T_1 to be generated from S_1 states. Since their (nonradiative) decay rate k_T is small, they accumulate until they totally hinder SE through a dipole-allowed efficient $T_1 \rightarrow T_2$ absorption (Triplet absorption), or through other nonradiative energy transfers (Singlet–Triplet or Triplet–Triplet Annihilation)

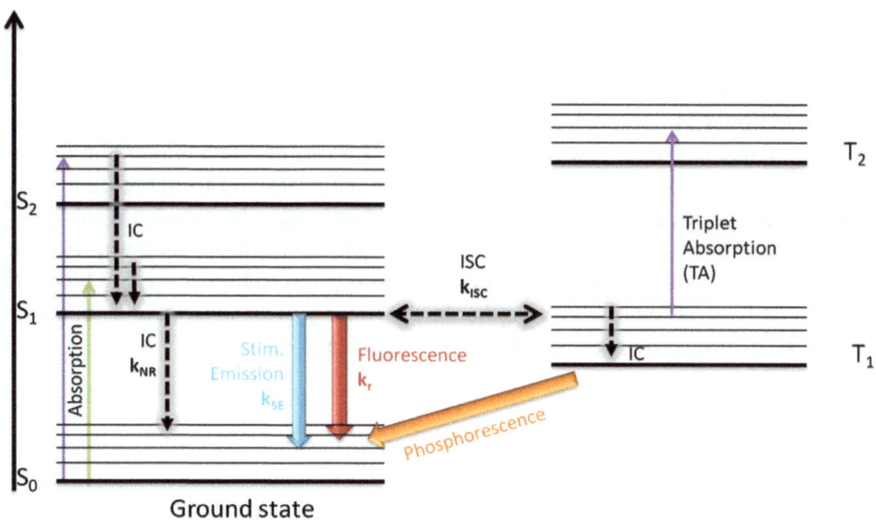

Fig. 2.11 A Jablonski diagram featuring the ground state (lowest vibrational level of the S0 manifold), the singlet S_1, S_2, … and triplet T_1, T_2, …excited states. Nonradiative decay occurs within the vibrational levels of a given electronic level, or from S_2 to S_1 for instance, through Internal Conversion (IC). Intersystem Crossing (ISC) occurs between S_1 and T_1

Triplet states play an especially important role in the understanding of organic lasers, and will be discussed with more details in Sect. 2.5.

Once a molecule has been promoted to the S_1 state, either directly from the ground state or after internal conversion (IC) from a higher S_n state, it first decays down to the lowest vibrational state by a rapid Internal Conversion process, which occurs within $\sim 10^{-14}$–10^{-10} s, then the molecule is deexcited through different possible mechanisms, briefly described below.

2.2.3.1 Fluorescence

The molecule can decay radiatively through **fluorescence**: the radiative decay is given by $k_r = \tau_{rad}^{-1}$ where τ_{rad} is the radiative lifetime, which is fundamentally connected to the absorption of the $S_0 \rightarrow S_1$ band through the Einstein relations. However as the latter apply strictly speaking to atomic systems with sharp lines, a modified formula is needed in practice to take into account the broad nature of absorption and emission: Strickler and Berg [39] derived the following useful relation connecting the radiative lifetime to the absorption cross section spectrum [40]:

$$k_r = \frac{1}{\tau_{rad}} = \frac{8\pi c n_F^3}{n_A} \frac{\int_{S_0 \rightarrow S_1(em)} f(\lambda) d\lambda}{\int_{S_0 \rightarrow S_1(em)} f(\lambda) \lambda^3 d\lambda} \int_{S_0 \rightarrow S_1(abs)} \frac{\sigma_{abs}(\lambda)}{\lambda} d\lambda$$

$$= \frac{8\pi c n_F^3}{n_A < \lambda^3 >_F} \int_{S_0 1(abs)} \frac{\sigma_{abs}(\lambda)}{\lambda} d\lambda$$

In this expression, n_A (resp. n_F) is the average refractive index of the material in the absorption (resp. emission) spectral region ($n_A > n_F$). The average of λ^3 appears denoted as $<\lambda^3>_F$, weighted by the fluorescence spectrum $f(\lambda)$; the latter integral is theoretically restricted to the $S_1 \rightarrow S_0$ emission region, but in most practical cases the whole emission spectrum is only ascribed to fluorescence from S_1 hence no special care has to be taken for the integration boundaries. Conversely, a higher care must be paid to the integral over the absorption spectrum, restricted to the $S_0 \rightarrow S_1$ band: it does not include the absorption to higher energy levels (S_n states) occurring at lower wavelengths. In general, $S_0 \rightarrow S_1$ and $S_0 \rightarrow S_n$ bands can be clearly distinguished in the absorption spectrum.

2.2.3.2 Internal Conversion

The molecule can decay nonradiatively directly to the ground state by **internal conversion (IC)**, with a nonradiative decay rate $k_{nr} = \tau_{nr}^{-1}$. A general rule of thumb relates the importance of nonradiative decay by IC to the structural rigidity of the molecule: as nonradiative decay requires some kind of vibrational coupling between the excited and the ground state, rigid and planar molecular structures will tend to favor luminescence versus IC, as illustrated in the highly fluorescent fluorescein molecule [16]. Internal conversion is favored when a resonance is found between a highly vibrational state of S_0 and the lowest-lying vibronic state of S_1: for this reason stretching vibrations associated to hydrogen atoms, which are highly energetic because of the low mass of the hydrogen atom, are likely to contribute to nonradiative decay (see e.g. the case of Rhodamine 110 [16]). Also for this reason long-wavelength emitters with a small energy gap are more likely to suffer from internal conversion (this effect is a result of the "energy gap law", see Sect. 2.2.5.) For green, blue or UV-emitting molecules however, a quantum yield of fluorescence <1 is rather attributed to ISC rather than IC (see below).

2.2.3.3 Intersystem Crossing

The molecule can transfer its excitation to the long-lived triplet state T_1 by **intersystem crossing (ISC)**, with a rate k_{ISC}. This phenomenon, which primarily depends on the extent of spin–orbit coupling in the molecule, is described with more details in the following (cf. Sect. 2.3.3).

Once a molecule has been transferred to the triplet state T_1, it may relax to the ground state with a rate $k_T = \tau_T^{-1}$ which is mostly a pure nonradiative decay because the T_1-S_0 transition is spin-forbidden. Radiative decay from the T_1 state is referred to as phosphorescence, it can be observed at low temperatures when vibrational excitations are frozen, or for molecules exhibiting a strong spin–orbit coupling. An emitter in the T_1 state may also be promoted to a higher-lying triplet state T_n by a spin-allowed transition, this is called **triplet absorption** (TA), with typical cross section $\sigma_{TT} \sim 10^{-16}\text{cm}^2$, or decay to the ground state in a bimolecular process through the encounter with a singlet state or a triplet state, by **singlet–triplet annihilation** (STA) or **triplet–triplet annihilation** (TTA) respectively. As triplet states and bimolecular phenomena play an important role in the physics of organic lasers, they are treated with more details in Sects. 2.5 and 2.6 of this chapter, respectively.

A useful figure-of-merit of organic emitters is the *quantum yield of fluorescence* $\phi_F (0 < \phi_F < 1)$. It is defined as the ratio of emitted fluorescence photons divided by the number of absorbed pump photons. This parameter is sometimes defined with additional quenching terms at the denominator (representing other nonradiative decay pathways for S_1 population), but it does not take into account stimulated emission as a possible decay path, i.e. it is always defined in a non-lasing case (I = 0).

$$\phi_F = \frac{k_r}{k_r + k_{nr} + k_{ISC}} = \frac{\tau_{nr}}{\tau_{rad} + \tau_{nr} + k_{ISC}\tau_{rad}\tau_{nr}} = \frac{\tau_F}{\tau_{rad}}$$

$\tau_F = \phi_F \tau_{rad}$ is the fluorescence lifetime, which is actually the measured or the effective lifetime of the singlet state, taking into account all possible sources of decay (except stimulated emission). The fluorescence lifetime is always shorter than the radiative lifetime.

2.2.3.4 Stimulated Emission

The molecule can decay radiatively through **stimulated emission** with a rate $k_{SE} = \sigma_{em} I$ where σ_{em} is the stimulated emission cross section (cm^2), and I the intracavity laser intensity expressed in photonic units (the photon flux per unit area).

2.2.4 Photophysical Parameters Relevant for Organic Lasing

2.2.4.1 The Saturation Intensity I_{SAT}

It is worth noting that the different downward decay rates depleting the S_1 state compete with each other, and among intramolecular phenomena the stimulated emission rate is the only one to be intensity-dependant. A useful quantity in laser physics is the laser saturation intensity defined as:

$$I_{sat}(photonic\ units) = \frac{1}{\sigma_{em}\tau_F}$$

$$or\ I_{sat}(energetic\ units) = \frac{hc}{\lambda_{laser}\sigma_{em}\tau_F}$$

Organic materials have saturation intensities which are of the order $I_{sat} \sim MW/cm^2$. I_{sat} is the (intracavity) laser intensity for which the stimulated emission rate $\sigma_{em} I$ is equal to the spontaneous decay rate $\frac{1}{\tau_F}$, the latter including all nonradiative decays by IC or ISC. Stated otherwise, $I > I_{sat}$ defines a lasing operating regime where stimulated emission is the dominant mechanism draining the S_1 population back to the ground state. In this regime, as every photon produced by stimulated emission "kills" one excited molecule to bring it back to the ground state, the population inversion becomes saturated and is very small, in the same (symmetric) fashion as the absorbance of a saturable absorber decreases when the incoming intensity surpasses the saturation intensity.

Let's clarify one point that is often misunderstood: the saturation intensity is an *intracavity* laser intensity and is not related directly to the pump laser intensity threshold; it is not either an indication of a "maximum" intensity that could be produced by a laser. A laser operating in a pure CW regime, let's say, is not an energy-storing device but an energy-converting device, converting the energy of the pump into a laser beam. If no additional effects come to complicate the 4-level diagram at high intensities, and if photodegradation issues do not exist, then in this perfect world we could convert the pump light into laser light with no upper limit in pump power, though with a less-than-one efficiency: the output/input intensity characteristic of the laser would be a straight line with a threshold and no upper bound.

Seeking the saturated regime ($I > I_{sat}$) is generally useful in laser systems where nonradiative pathways exist (and limit the efficiency and the photostability), because they can be quenched by stimulated emission. A high laser intensity may be obtained through the design of high-Q cavities for a given pump intensity. The reduction of the triplet formation channel by ISC has been observed when lasing was activated in a polymer laser [41]. The inhibition of spontaneous decay by stimulated emission in organic chromophores has also been exploited in a different context in microscopy, and is known as Stimulated Emission Depletion (STED) Fluorescence Microscopy [42].

2.2.4.2 Stimulated Emission Cross Sections

As organic media have very strong absorption cross sections, they have accordingly comparable stimulated emission cross sections and therefore exhibit optical gains which stand among the highest of all laser media (gain cross sections around 10^{-15} cm^2 are thus typically reported in conjugated polymers [43]) and are comparable with gains met in inorganic semiconductors. The linear gain g (usually expressed in cm^{-1}) is defined as:

$$\frac{dI}{dz} = g = \sigma_{em}\Delta N = \sigma_{em} S_1 N$$

where I is the laser intensity in photonic units (photon flux in m^{-2} s^{-1}) and ΔN is the population inversion (m^{-3}) between the upper level and the lower level of the lasing transition. As the latter level is rapidly flushed towards the ground state by internal conversion within picoseconds, the inversion is equal to the S_1 singlet state density. Here we denote as S_1 the relative population of the S_1 state, i.e. $\Delta N = S_1 N$, N being the molecular density. To fix ideas, with a chromophore density N of the order of $\sim 10^{19}$ cm^{-3} (dye-doped polymers, guest/host systems) to $\sim 10^{21}$ cm^{-3} (neat molecular solids, conjugated polymers), a complete inversion in an otherwise lossless medium would yield gains up to $\sim 10^4$–10^6 cm^{-1}.

These very high gains mean that the saturation intensity defined above can be reached in length scales that are of the order of a few mm at most, a limitation that is met when measuring gains by the variable Stripe length technique for instance (see Chap. 3).

Emission cross sections can theoretically be derived from optical gain characterizations (see Chap. 3), however it is not straightforward to measure independently the singlet state population, especially when steady state is not reached (which is often the case in organic lasers operating with short pulses). This is why it is preferred to use the following formula derived from the Strickler–Berg relation [1, 39]:

$$\sigma_{em}(\lambda) = \frac{\lambda^4}{8\pi c n_F^2 \tau_{rad}} \frac{f(\lambda)}{\int f(\lambda) d\lambda}$$

where n_F is the average refractive index in the emission spectral region, f(λ) the fluorescence spectrum (in arbitrary units because of the normalizing integral) and τ_{rad} is the radiative lifetime. Since the latter is connected also to absorption cross sections, it is possible to calculate the emission cross section from the absorption cross section spectrum only, by using the following relation

$$\int \frac{\sigma_{em}(\lambda)}{\lambda} d\lambda = \frac{n_F}{n_A} \int_{S_0 \to S_1(abs)} \frac{\sigma_{abs}(\lambda)}{\lambda} d\lambda$$

Or

$$\sigma_{em}(\lambda) = \frac{n_F}{n_A} \frac{\lambda^4 f(\lambda)}{\int \lambda^3 f(\lambda) d\lambda} \int_{S_0 \to S_1(abs)} \frac{\sigma_{abs}(\lambda)}{\lambda} d\lambda$$

One should keep in mind the comments done in Sect. 2.2.3 about the Strickler–Berg relation, regarding the integration boundaries. This last relation is particularly useful in practice as the absorption cross section spectrum, as explained in Sect. 2.2.1.3, is easily measurable (Table 2.1).

2.2.5 Short and Long Wavelength Limits for Organic Lasers

In this section we try to answer the following question: while it seems relatively "easy" to chemically tune the absorption/emission spectrum of an organic system, why is it so difficult to synthetize a molecule or a light-emitting polymer emitting efficiently enough in the ultraviolet or in the infrared to make it a laser material? In practice indeed, unless using wavelength-conversion tricks [44, 45], organic solid-state lasers turn out to be mostly destined to serve in the visible range.

On the short-wavelength side, deep-blue or UV organic emitters are difficult to find, although they recently triggered a great interest in virtue of their applications in organic light-emitting diodes, especially phosphorescent emitters [46]. As explained above (see Sect. 2.2.1.2), UV emitters will in general be characterized by a short conjugation length. There is no simple connection between the size of a molecular conjugation segment and the quantum yield of fluorescence: for instance the theoretical quantum yields of fluorescence in the acene series does not vary in a simple way starting from naphthalene (two benzene rings stuck together, deep UV emission, quantum yield 31 %) to pentacene (five benzene rings, emission in the red, quantum yield 23 %) [47], and some UV dyes like coumarin 1 emitting around 380 nm have quantum yields as high as 0.85 in oxygen-free solutions for instance [48]. As far as organic lasing is concerned, the most relevant issue about UV emitters is more practical in nature and deals with photostability. As a laser effect in the UV requires excitation with photons of higher energy in the UV or the deep UV (although there has been a demonstration of 2-photon pumping of a blue

Table 2.1 Summary of the main photophysical parameters used to describe the physics of an organic laser medium. For each parameter, a typical value, a typical range and an example are given

Photophysical parameter	Typical value	Typical range	Examples (material, reference)
Monomolecular (or intrachain) parameters			
Absorption cross section $\boldsymbol{\sigma_{abs}}$ @ peak abs. wavelength (cm^2)	10^{-16}	10^{-17}–10^{-15}	2 10^{-17} (MEH-PPV, [1]) 2.7 10^{-16} (Pyrromethene 567:copolymer P5MA-MMA, [29])
Fluorescence quantum yield Φ_f	70 %	0–100 %	30 % (MEH-PPV, [1]) 98 % (Coumarin 540:PMMA, [103])
Radiative lifetime $\boldsymbol{\tau_{rad}}$ (ns)	5	1–10	1.62 (Neat film Horner-type MEH-PPV, [1])
Emission cross section $\boldsymbol{\sigma_{em}}$ @ peak em. wavelength (cm^2)	10^{-16}	10^{-18}–10^{-15}	2 10^{-16} (Pyrromethene 567:copolymer P5MA-MMA, [29]) 1.1 10^{-16} (Horner-type MEH-PPV, [1])
Intersystem crossing rate $\mathbf{k_{ISC}}$ (s^{-1})	10^7	10^6–10^8 (10^{12a})	6.7 10^6 (F8BT, [104])
Triplet state T_1 lifetime $\boldsymbol{\tau_T}$ (s)	10^{-5}	10^{-3}–10^{-7} s	4.10^{-6} s (MEH-PPV, [62]
Triplet–triplet $T_1 \rightarrow T_n$ absorption cross section $\boldsymbol{\sigma_{TT}}$ (cm^2)	10^{-16}	10^{-17}–10^{-15}	2.10^{-16} (F8BT, [74])
Bimolecular (or interchain) parameters			
Singlet–singlet annihilation rate $\mathbf{k_{SS}}$ (cm^3/s)	10^{-9}	10^{-11}–10^{-7}	2.6 10^{-9} (F8BT, [105])
Singlet–triplet annihilation rate $\mathbf{k_{ST}}$ (cm^3/s)	10^{-9}	10^{-10}–10^{-8}	5.2 10^{-10} (Polyfluorene, [106])
Triplet–triplet annihilation rate $\mathbf{k_{TT}}$ (cm^3/s)	10^{-12}	10^{-14}–10^{-10}	10^{-14} (MEH-PPV, [107])
Singlet-polaron annihilation rate $\mathbf{k_{SP}}$ (cm^3/s)	10^{-9}	10^{-10}–10^{-8}	1.2 10^{-8} (PPV, [108])
Triplet-Polaron annihilation rate κ_{TP} (cm^3/s)	10^{-12}	10^{-13}–10^{-12}	10^{-12} (DCM2:CBP, [66])

[a] Values $\geq 10^{12}$ s^{-1} for intersystem crossing rates only apply for phosphorescent emitters containing heavy metal atoms (platinum or iridium organometallic complexes) which induce a very strong spin–orbit coupling. However, up to now these materials have not proven to be useful for lasing

polyfluorene laser [49]), this causes extremely fast degradation of the molecule, as high-energy photons have enough energy to cause direct bond breaking with creation of reactive radicals. Because of the large potential interest for spectroscopy, substantial efforts have however been done to look for stable and efficient UV emitters. Silafluorenes [50] or spiro-compounds [51] are good candidates for this purpose: the lowest lasing wavelength achieved to date directly from an organic semiconductor film is 361.9 nm [52], obtained with a thermally-evaporated spiro-terphenyl film.

In the infrared side, the reasons why it is hard to obtain emission from organic systems above ~700 nm are different and related to both an emission and an absorption issue. First of all, the emission from low-bandgap materials is fundamentally restricted by the "energy gap law" [53], an empirical law stating that the nonradiative decay rate of an excited state increases exponentially as the energy gap decreases. Telecommunications, biomedical applications (for instance, deep-tissue imaging) and probably in a near future plasmonics, will however keep on motivating research towards efficient deep-red or infrared gain materials and lasers. There have been recent reports of lasing operation from 890 to 930 nm [54], or even at 970 nm with a commercial dye (LDS 950) doped in a fluorinated polyimide waveguide [55]. The evidence of stimulated emission was brought through gain measurements at a wavelength as high as 1.3 μm in IR1051 dye [56] with a measured gain of 11 cm^{-1} under 1064-nm pumping.

But if emission meets this fundamental limit, we may wonder what happens for absorption bands: can we build *pi*-systems with bandgaps as low as we want, provided that we allow more and more bonds to be shared along a *pi*-conjugated segment? The question is worth asking, because long chains are met in conjugated polymers, and because building low bandgap organic materials is of considerable practical interest, for organic electronics or photovoltaics for instance, where maximizing the absorption of the red part of the solar spectrum is a long-sought objective [57]. If we take the very simple example of *trans* polyacetylene (see Fig. 2.12), which goes logically upon examination after the case of the smaller butadiene molecule considered in Sect. 2.2.1.2, one should expect the following scenario: as more and more π orbitals are shared, the gap between the HOMO and

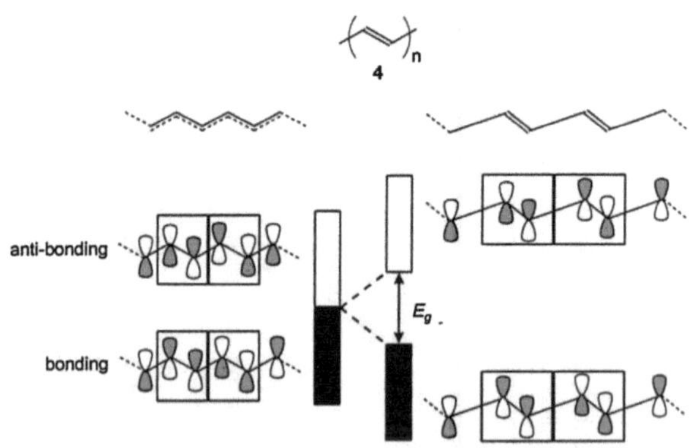

Fig. 2.12 Illustration of the "bond length alternation" principle in long trans-polyacetylene chains, and its consequences on the opening of a bandgap (Peierls instability). Courtesy of E.W. Meijer. Reprinted from [100] Copyright (2001), with permission from Elsevier

LUMO should decrease (see Fig. 2.6 for the case of oligothiophenes) until it eventually completely collapses to form a one-dimensional metal. This would indeed happen if the bond lengths between all carbon atoms were strictly identical and the electrons *fully* delocalized over the whole molecule; however this does not occur because this kind of equidistant linear chain structure happens to be unstable, an effect known as the *Peierls instability* [58]. The Peierls instability can be briefly understood as follows: consider that some extra energy is brought to the molecule which causes an elastic deformation, or a phonon, that results in alternating longer and shorter bonds; this excess of elastic energy stored by the molecule also produces a periodic potential for the *pi*-electrons. This periodic potential opens up a gap: this well-known fact (responsible for the electronic energy gap in crystals, for photonic gaps in photonic crystals, etc.) from Brillouin theory means that there are two wavefunctions of different energies which are associated with the same wavenumber: one wavefunction where maximal values (i.e. with maximum chance to find the electron at this place) stand between the longer bonds, and the other one having its maxima between the shorter bonds; of course the latter one has a lower energy since electrons are in average at shorter distance from the nuclei and are therefore more stabilized. It has been shown by Peierls [58] that the loss in the electronic energy of the molecule resulting from this gap opening is higher than the elastic energy required to generate it, meaning that the situation of "bond length alternation" is energetically favorable [59]. This is not systematically the case in all conjugated systems (in phenyl rings for instance the bond lengths are strictly equivalent), but here in the case of simple linear polyenic systems typically a gap of at least ~ 1.5 eV remains, which corresponds to a maximum wavelength for light absorption around 800 nm. This is the reason why it is generally difficult to find *pi*-conjugated polymers that are efficient in absorbing infrared radiation (and even worse in emitting IR beyond 1 μm). Synthetizing conjugated polymers with a low gap is useful for many purposes and is a hot topic of research: in the 80–90s it was more oriented towards the search for a zero-band-gap polymer capable of forming an intrinsically conductive material, namely a truly organic "synthetic metal" for energy storage or anti-static coatings [60]. Around the turn of the millennium, optoelectronics applications and organic photovoltaics became the driving forces of this research. However it must be noticed that as far as lasers are concerned, this lack of organic materials in the near IR is not problematic since it's a region where inorganic semiconductors and crystals already densely occupy the room.

2.3 Triplet States and Their Influence on Lasing

Although triplet states are somewhat elusive (they are usually not luminescent), they play a crucial role in organic laser physics. The harvesting of triplet excitons in phosphorescent organic light emitting diodes has been a real breakthrough in this area [61], allowing for a fourfold increase of their quantum efficiency; since then triplet states have been the subject of intense investigations in relation to the engineering of organic devices. As far as organic lasers are concerned, this starts from the understanding of the temporal dynamics in OSC lasers [41, 62, 63] and goes up to the development of advanced strategies for monitoring triplet decay, e.g. by adding to the active medium some "triplet scavengers" [64] or "triplet managers" [63], or through a clever use of metallic nanostructures acting as "plasmonic sinks" to remove the triplets [65]. Triplets also hold a central responsibility for the huge obstacles that stand on the road leading to the realization of an electrically-pumped organic laser diode [66–68]. For all these reasons triplets certainly deserve a particular attention in this chapter dedicated to the fundamentals of organic lasers. This chapter presents the basics of triplet states, but they will keep the top line in the next subsections dedicated to intermolecular phenomena and temporal dynamics of solid-state lasers. A comprehensive overview of triplet states in organic semiconductors can be found in the excellent review of Köhler and Bässler [69], from which many of the data presented in this section have been extracted.

2.3.1 Nature of Triplet States

A distinctive feature of organic semiconductors and more generally of all organic *pi*-conjugated systems used for lasing applications is the existence of well-defined spin states for the excited states. The spin of a state is given by the total spin of all electrons in all molecular orbitals; however electrons in filled orbitals are paired (one spin up, one spin down) so they contribute zero to the total spin. The spin is hence that of a two-electron system, one in the HOMO state and the other one in the LUMO state. The wavefunction describing the excited state can be written as a function of the two spatial coordinates $\mathbf{r_1}$ and $\mathbf{r_2}$ for each electron, and expressed as a product of a spatial wavefunction by a spin wavefunction:

$$\psi(r_1, r_2) = \psi_{spatial}(r_1, r_2)\psi_{spin}(r_1, r_2)$$

Two important principles of quantum mechanics have to be invoked here: first, the two electrons are indistinguishable, meaning that the probability density must verify $|\psi(r_1, r_2)|^2 = |\psi(r_2, r_1)|^2$. Secondly, the Fermi exclusion principle implies that the wavefunctions $\psi(r_1, r_2)$ and $\psi(r_2, r_1)$ cannot be the same, so they have to be of opposite signs $\psi(r_1, r_2) = -\psi(r_2, r_1)$ to be in accordance with the first relation: the total wavefunction has to be antisymmetric with respect to the exchange of the two electrons. This can be done in two ways: either $\psi_{spatial}(r_1, r_2)$ is symmetric and $\psi_{spin}(r_1, r_2)$ is antisymmetric, or the opposite. The wavefunctions of the four eigenstates of the two-particle system are summarized in Fig. 2.13. There is only one way to build an antisymmetric spin function for the two-electron system starting from linear combinations of one-electron spin states, this is the

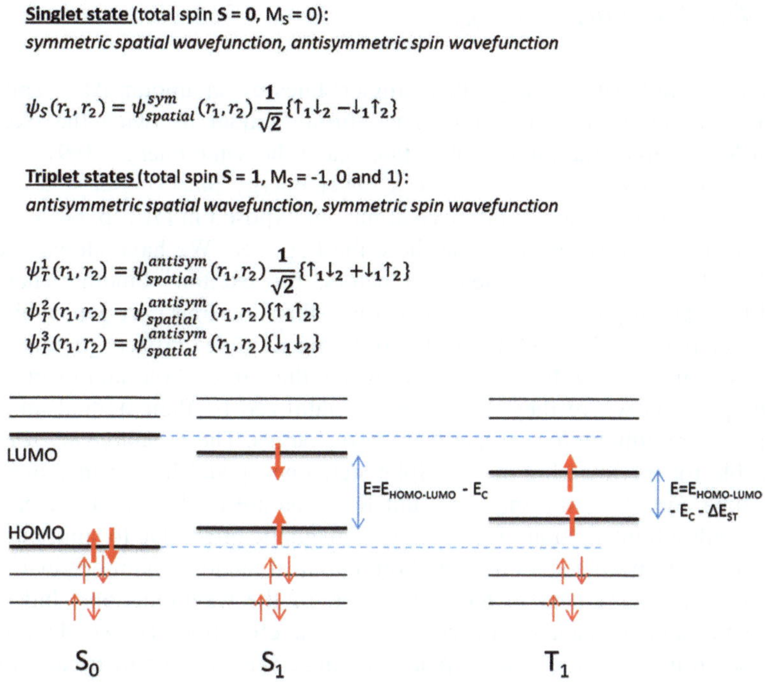

Fig. 2.13 *Top* singlet and triplet eigenstates wavefunctions. The ↑ and ↓ arrows denote the spin wavefunctions of the one-electron states with eigenvalues s = 1/2 and m_s = 1/2 and s = 1/2, m_s = −1/2 respectively. *Bottom* singlet and triplet states energy levels in an orbital configuration scheme. The S_1 and T_1 energy levels are shifted with respect to the HOMO–LUMO levels of the ground-state molecule by Coulomb and exchange interactions. The singlet state energy is given by $E_{S1} = E_{HOMO\text{-}LUMO} - J + K = E_{HOMO\text{-}LUMO} - E_c$, where E_c = J(coulomb integral) − K(exchange integral) is related to the exciton binding energy in an organic semiconductor; the Coulomb integral physically represents the coulomb attraction between the electron and hole in a pi-conjugated system. The triplet state energy is given by $E_{T1} = E_{HOMO\text{-}LUMO} - J - K = E_{HOMO\text{-}LUMO} - E_c - \Delta E_{ST}$, where $\Delta E_{ST} = 2$ K is the exchange energy. The exchange energy is an electrostatic term which reflects that the hole and the electron tend to be closer in a triplet state than in a singlet state because of the symmetry of the spatial part of the wavefunction (see text)

singlet state S_1, necessarily associated to a symmetric spatial wavefunction. In contrast, there are three symmetric spin eigenfunctions with a total spin S = 1, these are the triplet states.

Under electrical excitation, these four states have equal probabilities to be formed upon recombination of two spin 1/2 polarons (one hole and one electron), so the singlet formation yield is only 25 % in a fluorescent organic light-emitting diode.

2.3.2 The Exchange Energy

Triplet states lie lower in energy than singlet states by an amount ΔE_{ST} called the exchange energy or the singlet–triplet splitting (equal to twice the exchange integral K), whereas the three triplet states have the same energy[3] [69].

The reason why triplet states have a lower energy than singlet states can be understood from a physically-intuitive argument exposed in [70] (p. 68) or in [16] (p. 26), from which we only extract here the key idea. We have already seen in Sect. 2.1.2.1 that in organic semiconductors, the exciton[4] binding energy E_b caused the optical gap to be somewhat lower than the transport gap, which was itself lower than the HOMO–LUMO gap. Using a crude classical picture, this is because within the HOMO and LUMO bands the "free" hole and electron will seek the position where they are the more stabilized by their mutual attraction, making it more difficult to separate them into charges. The *exchange interaction* is responsible for the lowering of the triplet state energy and for the increase of the singlet state energy by the same amount K. Its nature is also purely electrostatic like the coulomb interaction [70], it is not a magnetic effect due to spin, but rather a pure quantum mechanical effect without classical analog which originates from the symmetry requirement of the spatial part of the wavefunctions. Indeed, for triplet states having an antisymmetrical spatial wavefunction, the two electrons can never be found at the same place because an antisymmetrical function $\psi_{spatial}(r_1, r_2)$ has necessarily a node for $r_1 = r_2$. Conversely, it is not forbidden for a singlet excited state with a symmetrical spatial wavefunction that electrons reside close to each other. With this crude picture in mind, we understand that electrons are more likely to be kept away from each other (or equivalently, for the hole and the electron to be kept closer to each other) in a triplet state than in a singlet state, hence leading to a triplet state having a lower potential energy.

From a quantum mechanical point of view, the exchange energy ΔE_{ST} scales exponentially with the overlap of the electron and hole wavefunctions [71]. Surprisingly, it was found to be around 0.7 eV in virtually all conjugated polymers, independently of the chemical structure, at the condition that the first excited state

---

[3] There is a small energy difference between the three triplet substates because of their different m_s quantum numbers, called the Zero Field Splitting (because it exists even in absence of an applied magnetic field) due to spin–spin interactions, however it is very low (from a few μeV to a few meV in metal–ligand charge transfer complexes) and is hence only detectable at cryogenic temperatures.

[4] The word *exciton* usually describes any *mobile* excited state, i.e. it is applicable whenever the excitation is able to travel or diffuse—to a nearby molecule for instance in the case of a molecular solid. The concept associated with the "exciton binding energy" is still valid for an oligomer or a dye dispersed in a non-conjugated matrix, although it may just be called the coulomb interaction in this case. Note also that the exciton binding energy in the case of an OSC carries more physical insight when it is defined with respect to the *transport gap,* as this becomes equal to the energy required to break the exciton into a pair of separated charges; in a dye-doped polymer the Coulomb energy can be expressed directly with respect to the HOMO–LUMO gap.

has a π–π^* character [69, 71]. In small molecules it may vary from ~0.1 to more than 1 eV. In Metal-to-Ligand Charge Transfer complexes for instance (mostly used in phosphorescent OLEDs), an electron is physically removed from the metal towards the ligand upon absorption of a photon, and this poor HOMO–LUMO overlap creates exchange energies as low as 0.2–0.3 eV [69]. In contrast, simple planar molecules like anthracene have an exchange energy as high as 1.5 eV [72]. The exchange energy is an important parameter as it is intimately connected with the Intersystem crossing rate (see below).

2.3.3 Optical Generation of Triplet States by Intersystem Crossing

The Wigner rule imposes that spin is conserved during an optical transition, meaning that only the S_1 state can be populated when a photon is absorbed from the ground state. This corresponds to a magnetic moment conservation rule, as the photon does not carry magnetic moment. However strict spin conservation during an optical transition would be true in a world where spin–orbit coupling does not exist. Spin–orbit coupling represents the magnetic mutual interaction between orbital momentum L and spin S: it is actually never zero (the spin–orbit coupling varies like Z^4 where Z is the atomic number), and it becomes especially strong when a heavy atom is present in the molecular structure [36, 73].

Spin–orbit interaction leads to a change in the selection rules, as L and S are not any more "good" quantum numbers to describe the system. It has two major consequences that are of interest for our subject: i) the T_1-S_0 transition becomes slightly allowed, the triplet state acquires a finite lifetime τ_T, though the latter remains several orders of magnitude longer than the singlet lifetime, typically in the μs to the ms range; and ii) a "spontaneous transfer" of the S_1 population towards the lower T_1 level becomes possible, called *intersystem crossing*, with a rate k_{ISC} defined by

$$\frac{dT_1}{dt} = k_{ISC} S_1$$

Intersystem crossing occurs because as the spin S is no longer a good quantum number, the spin states still referred to as "singlet" and "triplet" states are not in reality the true eigenstates of the system. For fluorescent emitters, ISC occurs within a typical timescale $1/k_{ISC}$ of the order of hundreds of ns ($k_{ISC} \sim 10^7 s^{-1}$). For phosphorescent emitters having a very strong ISC rate because of the presence of heavy atoms (Pt, Ir), the singlet state is not practically observable as any population in the singlet state very rapidly ends up in the triplet state with a timescale $1/k_{ISC} \sim$ ps.

The magnitude of k_{ISC} depends on the strength of the spin–orbit coupling and on the vibrational overlap between the initial and final wavefunctions. It follows the energy gap law for nonradiative transitions, i.e. it can be written as

$$k_{ISC} \propto exp(-A\,\Delta E)$$

where ΔE is the energy difference between the two states involved (and A is some constant). This means that high exchange energies are wanted to minimize intersystem crossing. This also means that the most efficient route for ISC will be first an horizontal transfer ($\Delta E = 0$) from the lowest vibrational state of S_1 towards a high-energy vibrational state of T_1, followed by a rapid vertical Internal Conversion (multiphonon decay) down to the lowest vibrational state of T_1, as sketched in Fig. 2.14, instead of a direct passage with $\Delta E = \Delta E_{ST}$. This picture of a horizontal transition followed by a very fast vertical decay is thought to be a correct general picture for nonradiative transitions between electronic states [69].

Even when no heavy atom is present in the molecule, some fluorescent dyes or polymers can have quite high intersystem crossing rates: the archetype for this is benzophenone, with $k_{ISC} = 10^{11}\ s^{-1}$ where the torsion of phenyl rings enables an efficient mixing of electronic states having different angular momenta.

2.3.4 Triplet Absorption

Once a molecule is in the T_1 state, it can absorb a photon toward a higher-lying triplet state, and because the transition is spin allowed, cross sections are high and typically of the same order of magnitude as singlet–singlet transitions ($\sigma_{TT} \sim 10^{-16}\ cm^2$). The first triplet state after T_1 that can be accessed optically is referred to as T_n, since the exact number of triplet states above is not usually known. Excitation to a T_n state will be generally followed by a fast nonradiative decay back to T_1, but this recovery is not complete: for instance, recovery experiments in the polymer F8BT have shown that 80 % of the T_n states relax back to T_1 rapidly (within 300 fs), while the remaining 20 % are supposed to end up in electron–hole pairs [74].

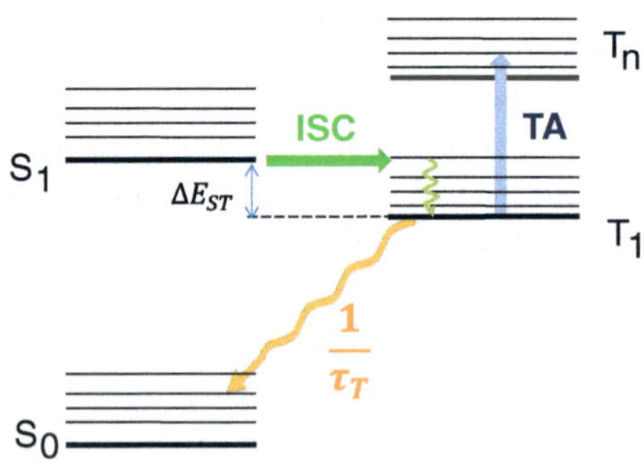

Fig. 2.14 Intersystem crossing (ISC), Triplet Absorption (TA) and exchange energy (ΔE_{ST}) in a conjugated system

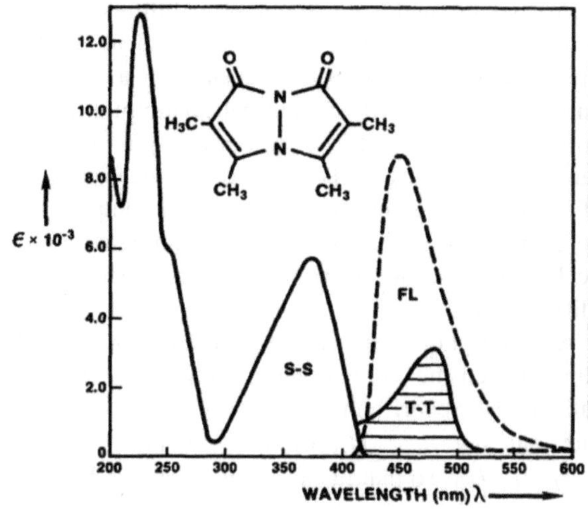

Fig. 2.15 Absorption (S–S), fluorescence (FL), and Triplet–Triplet (T–T) absorption spectrum of *syn*-(CH_3,CH_3)B in ethanol/ethyl ether 2:1 ratio. The T–T spectrum is recorded at 77 K. Reprinted with permission from [102]. Copyright 1986, American Institute of Physics

$T_1 \rightarrow T_n$ absorption bands are generally broad and significantly overlap with the fluorescence spectrum: it hence creates additional losses for the laser field that will be detrimental for efficiency, and will cause in the end the laser to shut down (see Sect. 2.5). In Fig. 2.15 is shown a typical Triplet–Triplet Absorption spectrum, evidencing the overlap between fluorescence (and hence gain) spectrum and T–T absorption.

There is a very intriguing fact about conjugated polymers with $\pi-\pi^*$ excited states (e.g. PPV derivatives or polyfluorenes): they happen to have a $T_1 \rightarrow T_n$ energy located around 1.5 eV, independently of the chemical structure. Because we already pointed out that the exchange energy was also similar, this means that a useful rule of thumb proposed by Köhler and Bässler [69] is to say that the T_1 state is 0.7 eV below the S_1 state, while the T_n state is 0.7 eV above the S_1 state for any conjugated polymer.

2.3.5 Phosphorescence: Are Triplet Emitters Suitable for Lasing?

Phosphorescence is the term used to describe luminescence from the triplet state T_1: while it can be present in any molecule at low temperatures, it becomes exploitable in a device when a very strong spin–orbit coupling is introduced by the presence of a heavy atom in the structure. Considerable efforts have hence been devoted in the first decade of the millennium to develop efficient phosphorescent emitters for Organic Light-emitting diodes, based on Metal-to-Ligand Charge Transfer complexes of Platinum or Iridium. The motivation for this research was driven by the fact that, under electrical excitation, the four possible exciton states resulting from the recombination of two spin-1/2 polarons of random spin (up or down) are created with equal probabilities, meaning that 75 % of triplets are formed, which are usually lost in fluorescent emitters. As the singlets end up very rapidly as triplets by ISC, electroluminescent devices based on phosphorescent emitters can reach 100 % internal quantum efficiency [75]. Good phosphorescent emitters must exhibit a very strong spin–orbit coupling to see their $T_1 \rightarrow S_0$ transition becoming frankly allowed. Green emitters are the most likely to be very efficient, since red emitters suffer from competition with nonradiative decay (it's the energy-gap law), and since blue emitters, associated to singlet states lying in the UV because of the exchange energy, suffer from very problematic photostability issues.

Are phosphorescent emitters suitable for lasing? Building a population inversion between the T_1 state and a high-lying vibrational level of the S_0 level is not difficult. However, considering a T_1 excited state, stimulated emission $T_1 \rightarrow S_0$ with rate $\sigma_{em(T_1 \rightarrow S_0)} \, IT_1$ directly competes with $T_1 \rightarrow T_n$ absorption with rate $\sigma_{TT(T_1 \rightarrow T_n)} \, IT_1$, so that lasing in an otherwise lossless medium will be possible at a wavelength λ at the simple condition that the stimulated emission cross section exceeds the absorption cross section at this wavelength:

$$\sigma_{em(T_1 \rightarrow S_0)}(\lambda) > \sigma_{TT(T_1 \rightarrow T_n)}(\lambda) \quad (\textit{condition of phosphorescent lasing})$$

This is a hard-to-meet requirement, because $T_1 \rightarrow S_0$ remains a forbidden transition in nature while $T_1 \rightarrow T_n$ is allowed, meaning that there is typically a three-orders-of-magnitude difference between the two cross sections, and unfortunately the two bands spectrally overlap significantly. A phosphorescent organic laser would be feasible if one is able to find a material where spin–orbit coupling is very strong, combined with a phosphorescence emission spectrum well shifted from the triplet absorption spectrum. Such a laser would be interesting since it would not experience the pulse duration limit encountered with singlet emitters, opening the way for a true CW laser. Accordingly, as triplets represent a huge obstacle in the route towards electrically-driven organic semiconductor lasers because they quench singlets or reabsorb the emitted laser light, achieving phosphorescent lasing would be an interesting achievement in this context too. Some attempts to observe Amplified Spontaneous Emission in highly-efficient phosphorescent emitters Ir(ppy)3 [60] or Btp2Ir(acac) [76] have been tried without success.

2.4 Intermolecular Phenomena: Quenching and Energy Transfer

A chromophore in a solid-state material is never alone. It is surrounded by a large assembly of identical chromophores and by foreign elements (chromophores of a different kind, dissolved oxygen or water or any other impurities, metallic electrodes, etc.) which all may interact with it and *quench* the emission of the chromophore.

Quenching is a very general term designating the different mechanisms through which a singlet or a triplet state loses its excitation upon interaction with a foreign "quencher". For light-emitting devices and particularly for lasers, quenching of emissive singlets is detrimental, whereas it may be conversely useful to quench the triplets as their accumulation causes absorption of the laser light.

Here we use the term "quenching" in its widest sense, as a loss of the excitation energy for a given emitter. This energy can be eventually lost as heat directly (the molecule switches to its own triplet state upon the influence of the quencher for instance), or transferred to another compound, where the energy may be transformed into heat as well, or into light if the acceptor is a chromophore. This is called *energy transfer*.

We will investigate the following phenomena:

- exciton diffusion and homo-FRET (Fluorescence Resonant Energy Transfer), which involve energy transfer from an excited fluorophore D* to a fluorophore D of the same kind *in the ground state*;
- Singlet–singlet annihilation, singlet–triplet annihilation or triplet–triplet annihilation (SSA, STA, TTA), involving two fluorophores of the same kind *both in the excited state* D*;
- energy transfers between a donor D and an acceptor A of different kinds. If A is a light-emitting unit, then we form a host/guest system, in which D absorbs light, transfers the excitation to A, which finally emits the light. Host/guest systems form a widely used class of materials useful for laser action (Table 2.2).

2.4.1 Physical Origin of Quenching/Energy Transfer

Let D be a "donor" chromophore. Let's examine how it may lose its energy by quenching. The rate equation for the excited state density (let's note it D* here) will be $\frac{dD^*}{dt} = -kD^*$. The transfer rate constant k depends on the distance r between D and the quencher (or acceptor) in two very different ways, which enable defining two types of mechanisms for quenching:

(a) **"Long-range" quenching/Resonant Energy Transfer RET** (or FRET for Fluorescence Energy Transfer or Förster Resonant Energy Transfer).

It is a nonradiative dipole–dipole interaction, which does not however proceed through the emission and absorption of a photon (Fig. 2.16). The transfer rate constant k_{RET} can be expressed as:

Table 2.2 Summary of some bimolecular processes in organic materials (see text for details)

Excited molecule		Other element					Process
D*	+	D	→	D	+	D*	Diffusion
D*	+	D*	→	D	+	D*	SSA/TTA
D*	+	Q	→	D			Quenching
D*	+	A	→	D	+	A*	FRET

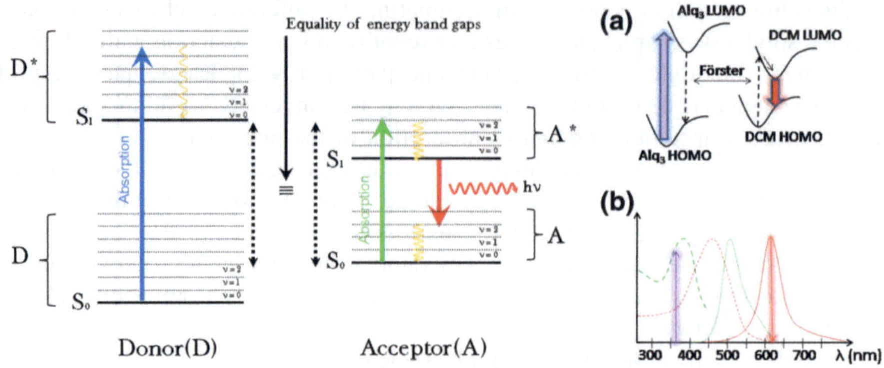

Fig. 2.16 Schematic representation of Förster resonant energy transfer (FRET). *Right* (**a**) Illustration of the host–guest Förster energy transfer between Alq_3 and DCM and (**b**) corresponding absorption (*dashed line*) and emission (*full line*) spectra. The Alq_3 host absorbs a UV pump photon and transfers its excitation through nonradiative dipole–dipole interaction to the guest; Förster energy transfer is efficient if the overlap between the emission spectrum of the host and the absorption spectrum of the guest is large. It results in a photon emission by the guest at a wavelength where the host absorption is almost zero, thus minimizing reabsorption losses

$$k_{RET} = \frac{1}{\tau_D} \cdot \left(\frac{R_0}{r}\right)^6$$

where τ_D is the donor lifetime (when no energy transfer occurs, ~ns), and R_0 is the Förster radius, defined as the distance for which spontaneous decay of the donor and energy transfer to the acceptor are equally probable. FRET is only possible between dipole-allowed transitions (S–S or T–T for instance but not S-T) and is efficient if the overlap between the emission spectrum of the donor and the absorption spectrum of the acceptor is high enough.

The Förster radius R_0 is typically a few nanometers, that is several times the size of a molecular unit: FRET is a distance interaction. It can be calculated from the following relation [77]:

$$R_0^6 = \frac{9000(\ln 10)\kappa^2 \Phi_D}{128\pi^5 N_A n^4} \int_0^\infty I_D(\lambda)\, \varepsilon_A(\lambda)\, \lambda^4 d\lambda$$

where Φ_D is the donor fluorescence quantum yield (when no transfer occurs), n is the average refractive index in the overlap spectral range, $I_D(\lambda)$ is the fluorescence spectrum of the donor (normalized so that the integral of I_D over λ is equal to unity), $\varepsilon_A(\lambda)$ is the acceptor molar absorption coefficient (in L mol^{-1} cm^{-1}), N_A is the Avogadro number and κ is an orientation factor (with κ^2 equal to 0.66 for free rotating molecules—in a liquid for example—and κ^2 equal to 0.476 for randomly dispersed acceptor molecules in a rigid medium—both in terms of relative distance and orientation).

The efficiency ρ of the energy transfer (or probability for D to transfer its excitation to A rather than decaying spontaneously) can be written as:

$$\rho = \frac{1}{1 + \left(\frac{r}{R_0}\right)^6}$$

The transfer efficiency is by definition equal to 50 % for $r = R_0$, and varies as $1/r^6$. A plethoric bibliography can be found about FRET as it is a widely used technique in biochemistry and spectroscopy [78].

(b) **Short-Range Quenching**

For this type of interaction, a physical contact between the donor molecule (the quenched unit) D and the quencher Q must exist, i.e. the electronic clouds must overlap. As the electron density decreases exponentially with distance from the nuclei, the rate of quenching depends on the distance as:

$$k_Q = A.exp[-\beta(r - r_c)]$$

where r is the distance between the donor and the quencher, r_c is the distance of closest approach at molecular contact, A is a constant ($\sim 10^{13}$ s^{-1}) and β is typically ~ 1 $Å^{-1}$ [77].

There are at least three mechanisms for short-range quenching:

- ***Intersystem crossing***: quenching of an excited state by molecular oxygen for instance is thought to occur by ISC: when O_2 comes in contact with a molecular unit, it accelerates its transfer to the triplet state where it becomes non-emissive [79].
- ***Photoinduced Charge Transfer***: an electron (or a hole) is physically removed from the donor D towards the acceptor A to form a (D^+ A^-) complex: this kind of quenching is a highly-wanted phenomenon in organic solar cells for instance, where it has to be more efficient than fluorescence of the absorbing material, which is in this case an unwanted recombination phenomenon.
- ***Dexter electron-exchange interaction*** (*or Dexter energy transfer*): in this process, an electron is exchanged from the LUMO of the donor to the LUMO of the acceptor, while simultaneously a hole is exchanged from the HOMO of the acceptor to the HOMO of the donor. The net result is a transfer of the exciton. Compared to Förster RET, Dexter energy transfer also requires a spectral overlap of the emission spectrum of the donor with the absorption spectrum of the acceptor, but in contrast D and A must be "in contact", and Dexter interaction does *not* require the transition to be dipole-allowed (hence Dexter transfer to/from triplet states is possible).

A summary of Förster versus Dexter interactions is presented in Fig. 2.17.

Up to now, to discuss the photophysics relevant for organic laser science, we did not pay much attention to differentiate neat organic solids (conjugated polymer

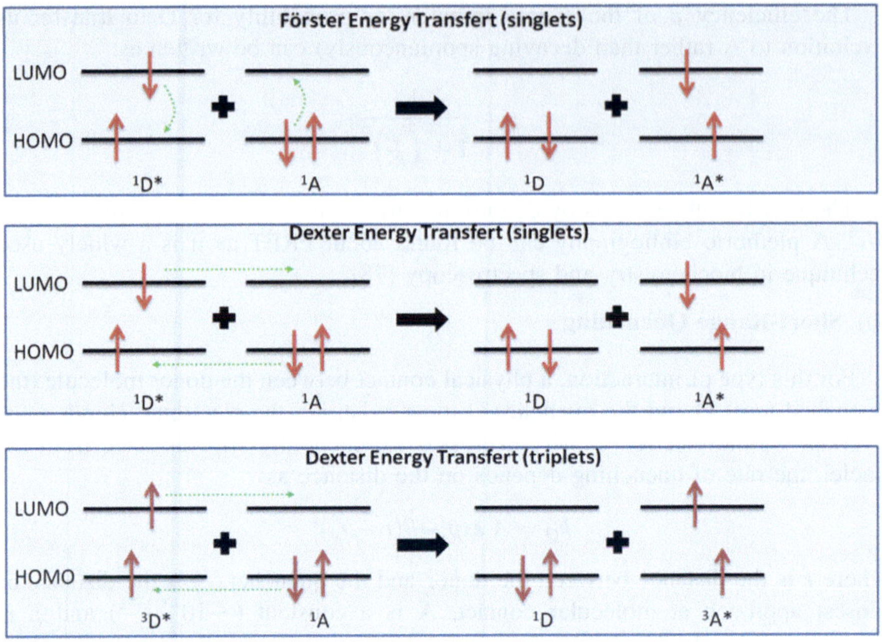

Fig. 2.17 Illustration of Förster energy transfer (*up*): it's a dipole–dipole interaction, formally equivalent to a photon emission by D which is reabsorbed by A (however no real photon is emitted); it does not require a contact between D and A, but an overlap between the D emission spectrum and the A absorption spectrum; furthermore the transition must be dipole-allowed, so that only singlet states can be concerned. For Dexter energy transfer, on the contrary, D and A electron clouds must overlap so that an simultaneous exchange of electrons and holes can occur. There is no restriction on selection rules and both triplets and singlets can intervene

films, molecular solids, molecular crystals) from dispersed emitters (dye-doped polymers). Now it is useful to remember that in a dispersed medium, only long range Förster RET is expected to occur between the active species (this may concern Singlet–singlet annihilation, singlet–triplet Annihilation, energy transfer to a ground state neighbor), to which one may add contact-type interactions but only with impurities or oxygen. Conversely, annihilation phenomena like Triplet–Triplet annihilation, or exciton diffusion are important in organic semiconductors but do not intervene in dispersed media.

2.4.2 Different Mechanisms of Bimolecular Interactions

Let's suppose we have a singlet excited state in a solid-state environment, that we shall note here D*, because upon quenching it will donate its excitation and act as a donor. Several scenarii are possible, listed in the introduction. We here focus on diffusion (and homoFRET), Singlet–Singlet annihilation (SSA), Singlet–Triplet annihilation (STA), and Triplet–Triplet annihilation (TTA).

2.4.2.1 Exciton Diffusion and HOMO-FRET

D* can be quenched by another identical molecule or emissive unit D in the ground state. The process is then called diffusion:

$$\boldsymbol{D}^* + \boldsymbol{D} \rightarrow \boldsymbol{D} + \boldsymbol{D}^* \textit{ exciton diffusion}$$

The transfer can proceed through a Dexter exchange interaction in neat films of OSCs, but may also arise when the two sites are away from each other, by FRET which is called in this case *homo-FRET* (only for singlets; the latter is efficient when there is a good overlap between absorption and emission spectrum, that is for a low Stokes shift). Diffusion is not directly a quenching loss mechanism, as the energy is just translated elsewhere, but it enhances the risk that the mobile excitation ends up in meeting a "real" killing quenching site. Exciton diffusion occurs within a typical exciton diffusion length $L_D \sim \sqrt{D\tau}$ where D is the diffusion constant and τ the exciton lifetime. Because triplets have lifetimes that are higher than singlet lifetimes, a typical order of magnitude for singlet diffusion length is a few nm, while for triplets it can reach several tens of nm or more [80, 81].

Homo-FRET is a non-negligible phenomenon in dispersed dye-doped polymer media even at low concentrations. Indeed, an active gain medium consisting of 1 % (in mass) of Rhodamine 640 molecules dispersed in a solid-state matrix of PMMA corresponds to a distance between adjacent molecules which is of the order of 3.3 nm if we assume a homogeneous blend, while the value of R_0 for the system {Donor = Acceptor = Rhodamine 640} is as high as 4.8 nm, because of a small Stokes shift for this molecule. As a consequence, the transfer efficiency in this case reaches 90 %. It falls at 50 % for a doping level of 0.4 % and is above 98 % for doping level higher than 2 %. HomoFRET can be evidenced by the depolarization of the fluorescence emission while increasing doping ratio [82].

2.4.2.2 Singlet–Singlet Annihilation

D* can be quenched by another identical molecule or emissive unit D* in the excited state:

$$\boldsymbol{D}^* + \boldsymbol{D}^* \rightarrow \boldsymbol{D}^* + \boldsymbol{D}$$

SSA is a process happening between two molecules in the first excited singlet state, i.e. when $\boldsymbol{D}^* = S_1$. During this phenomenon the energy from one of the excited molecules is transferred to the second one. As a result, the first molecule falls back to its ground state and the second molecule gets promoted to a higher excited singlet state S_n which then relaxes either to the first excited singlet state S_1 with a probability ξ or to the triplet state T_1 with a probability $1-\xi$.

Considering a probability ξ is a way to include intersystem crossing in the system in some way, and is consistent with the choice of some authors to treat ISC with a "yield" instead of a "rate", as we see in detail in Sect. 2.5.2.2. While some authors [83] do not take this effect into account and assume that all S_n states end up as S_1 ($\xi = 1$), some others [62, 84] assume that ξ is 0.25, which means that an excited singlet state produces on average one singlet for three triplets, exactly in the same proportions as when two polarons (one hole and one electron) encounter to make an exciton in an electrically-driven device [85].

For SSA [83] the reaction is sketched on Fig. 2.18 and writes:

$$S_1 + S_1 \to S_0 + S_n$$

Followed by

$$S_n \to \begin{matrix} S_1 + heat & (\xi) \\ T_1 + heat & (1-\xi) \end{matrix} \to S_0$$

The equation for the evolution of the S_1 population accounts for the fact that two singlets are lost, and only a fraction ξ of one of them is recovered, which writes:

$$\frac{dS_1}{dt} = -(2-\xi)k_{SS}S_1^2$$

k_{SS} is of the order of 10^{-9} cm^3/s (see Table 2.1) SSA is one of the only quenching phenomena that can explain the phenomenon of *concentration quenching* in dye-doped dispersed polymeric media, as it can arise through a long-range FRET mechanism provided that $S_0 \to S_1$ and $S_1 \to S_n$ bands spectrally overlap.

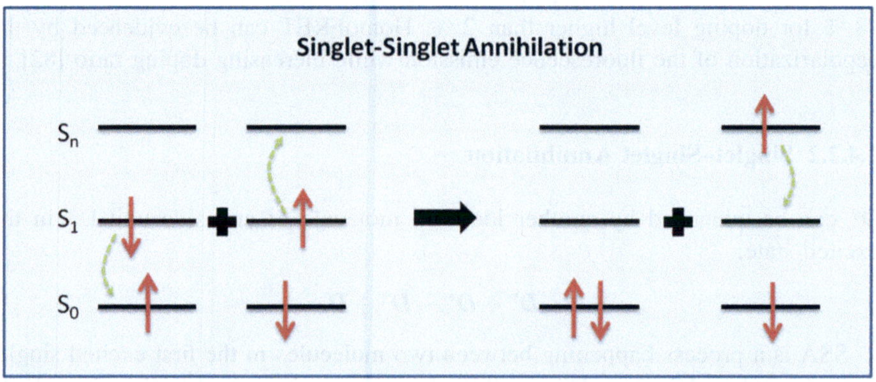

Fig. 2.18 Schematic representation of Singlet–Singlet Annihilation. Energy is transferred from one excited chromophore to a second excited chromophore resulting in a de-excitation of the first one along with excitation of the second chromophore to a higher excited state, from which it relaxes to the lowest excited state through fast non-radiative decay

Concentration quenching causes the quantum efficiency to drop dramatically for dye mass ratios inside the polymer above only a few percents [83].

2.4.2.3 Singlet–Triplet Annihilation

STA is a process where a molecule in the first excited singlet state S_1 is quenched by another molecule in the excited triplet state T_1 in its vicinity. As a result, the latter is getting excited to a higher triplet state while the former relaxes to its ground state. The molecule that is raised to an upper excited triplet state will then go down to the first excited triplet state which eventually relaxes to the ground state.

$$S_1 + T_1 \rightarrow S_0 + T_n$$

Followed by

$$T_n \rightarrow T_1 + heat \rightarrow S_0$$

Here all excited-state triplets end up in the T_1 state, as reverse ISC is a very unlikely phenomenon. The rate equation for STA is then:

$$\frac{dS_1}{dt} = -k_{ST} S_1 T_1$$

k_{ST} is of the order of 10^{-9} cm^3/s (see Table 2.1). The simultaneous transitions $S_1 \rightarrow S_0$ and $T_1 \rightarrow T_n$ are both allowed meaning that STA can arise through a Förster-type reaction. List et al. have shown that FRET is indeed the dominant mechanism for STA in conjugated polymers [86].

2.4.2.4 Triplet–Triplet Annihilation

TTA has been the subject of many investigations and debates [69]. It plays a major role in the triplet dynamics in OSCs and especially for devices under electrical excitation, because triplets are abundant (75 % of excitons created by electron/hole pairs recombinations) and long-lived, so that their mutual destruction by TTA

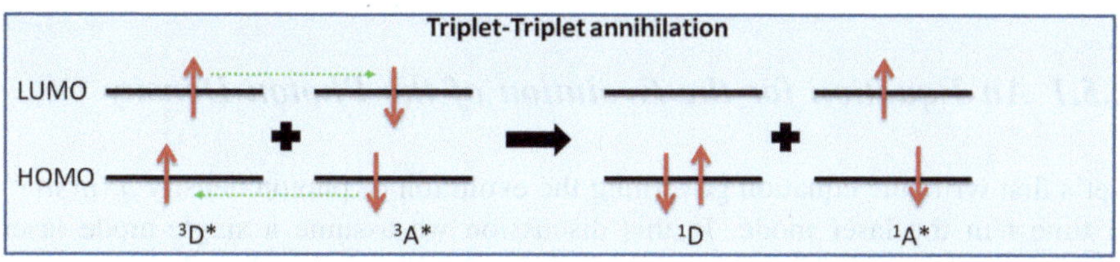

Fig. 2.19 Illustration of Triplet–Triplet annihilation (here occurring through a Dexter or exchange mechanism)

turns out to be an important issue. In the context of organic lasers under optical excitation, it is somewhat less important, and can be considered as a positive quenching phenomenon, as it helps depleting the triplet population and creating singlets out of triplets. TTA is incorporated in some organic lasers models [62, 84] (Fig. 2.19).

Because T-S transition are spin-forbidden, TTA can only be explained by a Dexter or exchange interaction, and simple images such as the one displayed in Fig. 2.18 for SSA fail at explaining what happens in a simple way [87]. TTA reaction can be written as [84]:

$$T_1 + T_1 \rightarrow S_1 + S_0(\xi) \quad \textit{and} \quad T_1 + T_1 \rightarrow T_1 + S_0(1 - \xi)$$

which is (at least formally) equivalent to two T_1 states reacting towards a high-lying S_n singlet state, which afterwards decays either towards S_1 with a probability ξ or towards T_1 with a probability $1 - \xi$, in a similar way as ξ was defined for SSA (see Sect. 2.4.2.2). The formation of singlet excitons S_1 from the annihilation of two long-lived triplet states is demonstrated experimentally through the observation of delayed luminescence. The rate equation for T_1 in presence of TTA is

$$\frac{dT_1}{dt} = [-2 + (1 - \xi)]k_{TT}T_1^2 = -(1 + \xi)k_{TT}T_1^2$$

k_{TT} is of the order of 10^{-12} cm^3/s (see Table 2.1).

2.5 Equations for Organic Solid-State Lasers

In this section we derive a few simple equations that can serve as a basis for theoretical work with organic lasers. Triplets hold the key role here, because the piling up of these long-lived states hinders lasing in the CW regime.

Organic lasers can hence only emit *short pulses*, unless specific action is taken to avoid the triplet problem. A simple solution was found with liquid dye lasers which can run in CW mode because the continuous circulation of the fluid allows refreshing the medium permanently. Another option is to reduce the triplet lifetime by adding "triplet quenchers", that is molecules that will interact preferentially with triplets to bring them back to the S_0 state through internal conversion.

2.5.1 An Equation for the Evolution of the Photon Density

Let's first write the equation governing the evolution of photon density q (in m^{-3}) at time t in the laser mode. In this discussion we assume a single-mode laser. Photons can be created by spontaneous or stimulated emission, but also reabsorbed or coupled out of the cavity to build up the useful laser light beam. This writes:

$$\frac{dq}{dt} = \left(\frac{dq}{dt}\right)_{stim} + \left(\frac{dq}{dt}\right)_{spont} + \left(\frac{dq}{dt}\right)_{material\,losses} - \frac{q}{\tau_{cav}}$$

The photon density has two feeding terms, spontaneous and stimulated emission, and two loss terms; The losses are of two different natures: the losses due to the material intrinsic photophysics (reabsorption of laser light, triplet–triplet absorption), referred here to as *material losses*, and the losses due to the cavity. The latter are the outcoupling losses of the mirrors, for an external cavity, or the losses embedded in the laser resonator, e.g. scattering or diffraction losses for a Distributed Feedback laser, for instance. Cavity losses can be written as $-\frac{q}{\tau_{cav}}$ where τ_{cav} is the *photon lifetime in the cavity*. The cavity lifetime physically represents the dwelling time of a photon injected into an optical cavity before it comes out. It will be of course high when long cavities with high reflectivity mirrors are used. In the case of a simple linear cavity of length L with two mirrors of reflectivities R_1 and R_2 facing one in front of each other enclosing a medium with a single-pass absorption A ($0 < A < 1$), it is found to be [88]:

$$\tau_{cav} = \frac{-2L}{cln\left[R_1R_1(1-A)^2\right]}$$

The term $\left(\frac{dq}{dt}\right)_{stim}$ accounts for stimulated emission (as well as for reabsorption from the laser transition lower level, but since the latter is quickly depopulated towards the ground state, this reabsorption is usually neglected) and writes:

$$\left(\frac{dq}{dt}\right)_{stim} = \Gamma\sigma_{em}I\Delta N = \Gamma\sigma_{em}\frac{c}{n_{eff}}qS_1$$

where the intracavity laser intensity I relates to q through $I = \frac{c}{n_{eff}}q$ (n_{eff} is the effective refractive index of the laser mode), ΔN is the population inversion density, equal to the S_1 singlet state population density.

Γ is the confinement factor defined by

$$\Gamma = \frac{\int_{active\ layer}|\langle\Pi(x)\rangle|dx}{\int_{-\infty}^{\infty}|\langle\Pi(x)\rangle|dx}$$

where $\langle\Pi(x)\rangle$ is the time-averaged Poynting vector (the power flow of the optical wave) in a gain medium consisting of an active layer confining light in the x-direction, and otherwise infinite in y- and z- directions (see Fig. 2.20). For external cavities or vertical geometries, the confinement factor is equal to $\Gamma = \frac{volume\ of\ the\ active\ medium}{laser\ mode\ volume}$; it can be well below unity and is in this case given by the ratio d/L of the gain medium thickness over the cavity length.

The term $\left(\frac{dq}{dt}\right)_{spont}$ accounts for spontaneous emission: it is the "feeding term" or the "trigger" in this equation which will provoke the avalanche of photons through the stimulated emission process. Although spontaneous emission

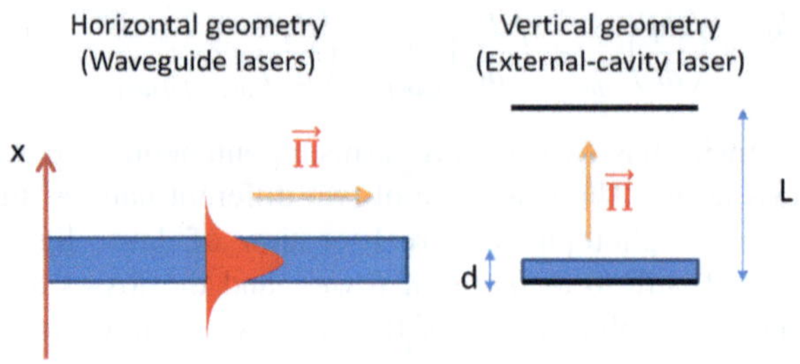

Fig. 2.20 Definition of the confinement factor: $\frac{\int_{active\ layer} |\langle \Pi(x) \rangle| dx}{\int_{-\infty}^{\infty} |\langle \Pi(x) \rangle| dx}$ where $\langle \Pi(x) \rangle$ is the time – average Poynting vector. In a horizontal geometry the light propagation direction (or the Poynting vector $\vec{\Pi}$ representing the power flow of the optical wave) is orthogonal to the confining direction x of the waveguide; the confinement factor is hence close to 1. In a vertical geometry with an external mirror (VECSOL case) the Poynting vector is along the thin-film direction and is just equal here to $\sim$d/L < 1

contributes generally in a negligible way to the photon density in the laser mode while the laser is operating, it is essential to give to this term a correct value to start the process. However, mistaking about this term is not dramatically important, except maybe if we are interested in laser behavior near threshold or lasers operating below the saturation intensity.

Many authors [62, 65, 67, 89] chose to write this term as $\left(\frac{dq}{dt}\right)_{spont} = \Gamma \beta_{spont} \frac{S_1}{\tau_F}$ where β_{spont} represents the fraction of all spontaneous photons (produced at a rate $\Gamma \frac{S_1}{\tau_F}$) emitted in the mode volume. In the reports cited above and quite generally in the literature, following [90] β_{spont} is taken equal to $\sim 10^{-4}$; however this term depends on the specific geometry of the laser resonator, and can become significantly higher in micro or nanocavities where only few optical modes are supported. In the general case, we may use the following result [91]: "*the spontaneous emission rate into one mode is equal to the rate of stimulated emission induced by one photon in this particular mode*". Let I_1 be the intensity of one photon in a mode volume V_{mode}, then $I_1 = \frac{c}{V_{mode}}$, and

$$\left(\frac{dq}{dt}\right)_{spont} = \Gamma \sigma_{em} I_1 \Delta N = \Gamma \sigma_{em} \frac{c}{V_{mode}} S_1$$

Then we obtain this simple relation for β_{spont} : $\beta_{spont} = \sigma_{em} I_1 \tau_F$.

At last, the term $\left(\frac{dq}{dt}\right)_{material\ losses}$ accounts for ground-state reabsorption at the laser wavelength, excited-state absorption ($S_1 \rightarrow S_n$) and Triplet–Triplet absorption (and other relevant terms if any, depending on the specific nature of the system).

$$\left(\frac{dq}{dt}\right)_{material\ losses} = -\sigma_{abs}^{S_0\rightarrow S_1}(\lambda_{laser})S_0 I - \sigma_{abs}^{S_1\rightarrow S_n}(\lambda_{laser})S_1 I - \sigma_{abs}^{T_1\rightarrow T_n}(\lambda_{laser})T_1 I$$
$$= -\alpha_{reabs} I = -\alpha_{reabs}\frac{c}{n_{eff}}q$$

In conclusion, we have this final formula governing the evolution of the intracavity photon density:

$$\frac{dq}{dt} = \left[\Gamma\sigma_{em}\mathrm{S}_1 - \alpha_{reabs} - \frac{n_{eff}}{c\tau_{cav}}\right]\frac{c}{n_{eff}}q + \Gamma\sigma_{em}\frac{c}{V_{mode}}\mathrm{S}_1$$

This equation couples the photon density q to the population inversion S_1. Rate equations will give us the missing relations in order to solve completely the problem.

2.5.2 Rate Equations: A Set of Equations Governing the Flow between Different States

Organic lasers are properly described by *rate equations* governing how the populations in the different states flow from a state to another under the effects of absorption, spontaneous emission, stimulated emission or nonradiative decays (internal conversion and intersystem crossing). However taking into account every vibrational level, or even every electronic level susceptible to intervene in the laser process is an impossible and irrelevant task. Equations are not more complicated when many higher excited states are involved to account for excited-state absorption or bleaching for instance, and models incorporating as many states as wanted can be easily implemented and solved numerically (even though a complex model with many unknown parameters will of course not be very helpful in capturing the physics of a given device).

Rate equations can be as complex as desired, if all quenching effects and absorption/emission to higher-lying excited states are considered. Incorporating all bimolecular phenomena (SSA, TSA, TTA) discussed in Sect. 2.4.2 as well as ISC enable writing for S_1 and T_1 the following set of equations:

$$\begin{cases} \frac{dS_1}{dt} = \sigma_{abs}I_pS_0 - \left(\frac{1}{\tau_F} + \sigma_{em}I\right)S_1 - k_{ST}S_1T_1 - (2-\xi)k_{SS}S_1^2 - k_{ISC}S_1 \\ \frac{dS_0}{dt} = -\sigma_{abs}I_pS_0 + \left(\frac{1}{\tau_F} + \sigma_{em}I\right)S_1 + k_{ST}S_1T_1 + \frac{T_1}{\tau_T} + k_{SS}S_1^2 + (1+\xi)k_{TT}T_1^2 \\ \frac{dT_1}{dt} = k_{ISC}S_1 - \frac{T_1}{\tau_T} + (1-\xi)k_{SS}S_1^2 - (1+\xi)k_{TT}T_1^2 \end{cases}$$

Here S_1, T_1 and S_0 are the normalized densities ($\mathrm{S}_1 + \mathrm{T}_1 + \mathrm{S}_0 = 1$); $\sigma_{abs}, \sigma_{em}$ are the absorption/emission cross sections (m^2) resp. at the pump/laser wavelength, I_p and I are resp. the pump and the intracavity laser intensity expressed in photonic units (photons/m^2/s); τ_F and τ_T are the fluorescence (singlet) lifetime and the triplet lifetime, respectively.

Here we also incorporated the equation for S_0, following the assumption that $S_1 + S_0 + T_1 = 1$. This assumption neglects all sources of permanent photodegradation (the total number of molecules is constant), and is essentially useful at high pumping intensities when the ground state can be depleted. Most of literature reports make a stronger approximation, assuming that only low inversions are obtained, i.e. $S_1, T_1 \ll 1$ (which is certainly true in OSCs but is questionable in dispersed media where the chromophore density is low and absorption saturation regime may be attained); this means that S_0 is taken as a constant equal to 1, and the pumping term hence becomes described by a *constant pump rate* $G = \sigma_{abs} I_p$ (s^{-1}). With this assumption, and using the photon density $\frac{n_{eff}}{c} I = q$ instead of the intracavity intensity, we have the following set of equations [62, 67, 92]:

$$\begin{cases} \frac{dS_1}{dt} = G - \left(\frac{1}{\tau_F} + \sigma_{em} \frac{c}{n_{eff}} q\right) S_1 - k_{ST} S_1 T_1 - (2 - \xi) k_{SS} S_1^2 - k_{ISC} S_1 \\ \frac{dT_1}{dt} = k_{ISC} S_1 - \frac{T_1}{\tau_T} + (1 - \xi) k_{SS} S_1^2 - (1 + \xi) k_{TT} T_1^2 \end{cases}$$

2.5.2.1 The 4-Level Approximation

In absence of quenching (i.e. STA, SSA, TTA, etc.) and if we totally neglect the influence of triplet states, the laser acts as a four-level system. This is a relevant approximation in experimental conditions where *pulsed sources with durations much smaller than the ISC time constant (on the order of hundreds of ns) are used.* Triplet accumulation and the subsequent phenomena are hence negligible. In addition, emission reabsorption from S_1 to a higher excited singlet state and stimulated emission at the pump wavelength are also neglected. The equations then simply write:

$$\begin{cases} \frac{dS_1}{dt} = \sigma_{abs} I_p S_0 - \left(\frac{1}{\tau_F} + \sigma_{em} I\right) S_1 \\ \frac{dS_0}{dt} = -\sigma_{abs} I_p S_0 + \left(\frac{1}{\tau_F} + \sigma_{em} I\right) S_1 \end{cases}$$

where I_p is the pump intensity, I the intracavity laser intensity, expressed in photonic units.

2.5.2.2 Taking Triplet States Into Account: The "Constant Yield" Versus "Constant Rate" Models

The 4-level model can be useful for many purposes but fails at capturing most of the interesting physics with organic materials which rely on the existence of triplet states or bimolecular interactions. Triplet filling is a well-known effect which generally hinders lasers from operating in the CW regime.

There has been intense research to determine how triplet states are formed under optical as well as electrical excitation [69, 93], and triplets have become extremely studied with the advent of triplet-harvesting phosphorescent organic light-emitting diodes [61].

The most widespread picture, used for decades in liquid dye lasers, consists in assuming that all triplets are created through intersystem crossing from the first singlet excited state S_1 with an intersystem crossing rate k_{ISC} (s^{-1}). The rate equations in a "constant triplet formation rate" or "k_{ISC} model" write:

$$\begin{cases} \frac{dS_1}{dt} = \sigma_{abs} I_p S_0 - \left(\frac{1}{\tau_F} + \sigma_{em1} I + k_{ISC}\right) S_1 \\ \frac{dS_0}{dt} = -\sigma_{abs} I_p S_0 + \left(\frac{1}{\tau_F} + \sigma_{em} I\right) S_1 + \frac{T_1}{\tau_T} \\ \frac{dT_1}{dt} = k_{ISC} S_1 - \frac{T_1}{\tau_T} \end{cases}$$

The constant rate model has been used for most works on solid-state dye-doped polymers or liquid dye lasers [16, 94, 95] and also in many recent works devoted to the dynamics of organic semiconductor films under optical excitation [62, 63, 96]. Conversely, another picture used in some papers [65, 67, 93, 97] consists in assuming that for pump photon excitation energies above that of the singlet level, triplets may be formed not only by ISC but also through exciton fission, fusion, recombination of photogenerated polarons or any mixture of phenomena which is then described empirically by a *constant yield of triplet formation* ϕ_T instead of a constant ISC rate. Each absorbed photon will produce a triplet state with a probability ϕ_T and a singlet state with a probability $1 - \phi_T$. The equations are as follows:

$$\begin{cases} \frac{dS_1}{dt} = \sigma_{abs} I_p (1 - \phi_T) S_0 - \left(\frac{1}{\tau_F} + \sigma_{em} I\right) S_1 \\ \frac{dS_0}{dt} = -\sigma_{abs} I_p S_0 + \left(\frac{1}{\tau_F} + \sigma_{em} I\right) S_1 + \frac{T_1}{\tau_T} \\ \frac{dT_1}{dt} = \sigma_{abs} I_p \phi_T S_0 - \frac{T_1}{\tau_T} \end{cases}$$

The differences in the models appear in Fig. 2.21.

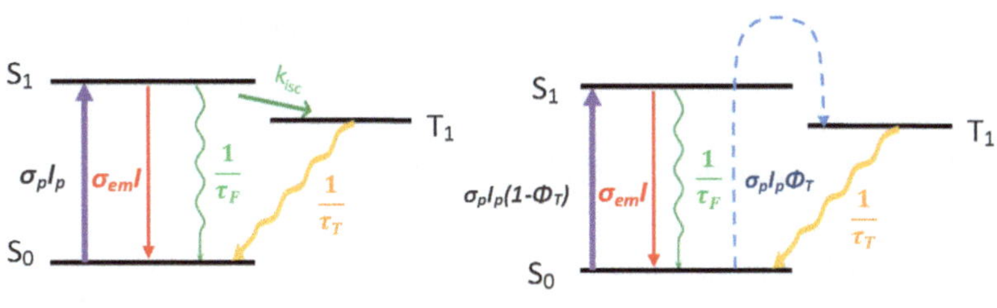

Fig. 2.21 Illustration of the "constant intersystem crossing rate" picture (*left*) versus the "constant triplet yield" picture (*right*) used to describe how triplets are formed in organic pi-conjugated systems under optical excitation

Although there is actually not so much difference in using either one or another model as soon as stimulated emission is absent, the models strongly differ in the presence of stimulated emission, this is why a discussion on these two models is especially relevant when dealing with organic lasers, and may not be that important in other contexts. Indeed, when I = 0 the two sets of equations are equivalent upon writing:

$$\phi_T = \frac{k_{ISC}\tau_F}{1 + k_{ISC}\tau_F}$$

In general, the intersystem crossing time constant $1/k_{ISC} \sim 10^{-7}$ s is much longer than the fluorescence lifetime $\tau_F \sim 10^{-9}$ s, so that the triplet yield is simply $\phi_T \approx k_{ISC}\tau_F$ (~ 1 %).

Solving the above rate equations in the steady-state is straightforward (all derivatives are set to zero) and enables picking up the main physical difference between the two models: in the "constant triplet yield" approach, the steady-state triplet population T_1 increases both with the pump intensity and with the laser intensity I; on the contrary in the "constant rate" approach, ISC is one of the channels which can be activated to deplete the singlet state (with fluorescence, IC, SE…): this means that if stimulated emission becomes strong enough ($I > I_{sat}$) then the feeding rate of the triplet level vanishes and T_1 hence decreases when the laser intensity I increases, for a fixed pump intensity.

2.6 Temporal Dynamics of Organic Solid State Lasers

Recent reports have shown the termination of lasing after a few ns [62] in polymer guest/host systems (see also Fig. 2.23) or a few tens of ns [67] in small-molecular lasers after pump turn-on in case of long-pulse pumping: these aspects are also essential in the context of electrical pumping. The great majority of optically pumped solid-state laser are operating with fs to ns pulse duration, limiting the effect of triplet states.

In this section, we derive a simple design formula for the pulsewidth and a condition for CW lasing, and we present afterwards the results of a numerical simulation from the above-exposed equations in a simple case.

2.6.1 The CW Lasing Condition

True CW lasing in organic solid-state lasers (that is, without rotating the medium to mimic the flow of a liquid dye laser, for instance) is a long-thought objective that has not been achieved yet at the time of writing this book [63].

In order to derive a simple useful analytical CW lasing condition, we first consider the previous rate equations within the framework of the "constant rate" model, and solve them at steady state (d/dt = 0). The equation for the triplet state reads

$$T_1^{ss} = k_{ISC}\tau_T S_1^{ss}$$

where T_1^{ss} and S_1^{ss} are the steady-state population of the T_1 and S_1 state, respectively. In order to have a rough estimate, we can assume that the only cause of loss in the system is Triplet Absorption. We neglect STA for instance, and we don't take into account other losses in the cavity. CW lasing will then be possible whenever triplet absorption probability will be lower than stimulated emission probability, or

$$\sigma_{TT}T_1^{ss} < \sigma_{em}S_1^{ss}$$

or

$$\sigma_{TT}k_{ISC}\tau_T < \sigma_{em}$$

simplified condition for CW lasing in an organic laser

It is clear from the typical orders of magnitude exposed in Table 2.1 that this condition is *not met* in most organic systems (σ_{em} and σ_{TT} have roughly the same order of magnitude but k_{ISC} is much higher than $1/\tau_T$, the triplet state is filled much more rapidly than it is depleted).

A good material for CW lasing should then have a low triplet absorption cross-section, a low ISC rate and a small triplet lifetime to allow the triplet population to pour out quickly. Furthermore the triplet absorption spectrum should be shifted as much as possible away from the emission spectrum to enable the previous relation to hold at a given wavelength within the emission spectrum, relaxing the very hard condition on ISC rate and triplet lifetime.

If we look for the CW lasing condition with the *constant yield* approach, we may assume that pumping occurs at a constant rate $G = \sigma_{abs}I_pS_0$ (no ground state depletion), and without any assumption on compared dynamics of triplets versus singlets, the condition $\sigma_{TT}T_1^{ss} < \sigma_{em}S_1^{ss}$ translates into a relation where the CW laser intensity I_{CW} intervenes:

$$\frac{\sigma_{em}\tau_F}{\sigma_{TT}\tau_T}\left(\frac{1-\phi_T}{\phi_T}\right) > 1 + \frac{I_{CW}}{I_{sat}}$$

With $I_{sat} = \frac{1}{\sigma_{em}\tau_F}$. Not surprisingly this relation is identical to the "constant rate" formulation $\sigma_{TT}k_{ISC}\tau_T < \sigma_{em}$ when $I_{CW} = 0$, which is the limit where the constant rate and constant yield models merge. The condition on σ_{TT} appears to be harder to meet when I_{CW} becomes comparable or higher than I_{sat}: this is physically because if the intracavity intensity is high, the singlet state is depleted, which is not the case of the triplet state, as the latter is continuously filled by the pump at a constant rate $\phi_T\sigma_{abs}I_pS_0$ whatever the laser intensity.

Comparing these models for CW laser action remains highly speculative; it is however useful to have in mind the simple rule-of-thumb given by the constant ISC model which is an adequate and simple rule to work with.

Recently, Zhang and Forrest [63] have published a theoretical paper predicting that in some optically pumped organic semiconductor lasers based on guest/host systems, two threshold pump intensities exist; one for pulsed lasing, I_{PS}, and another for continuous-wave (CW) lasing, I_{CW}. The theory presented in this paper predicts a decrease in I_{CW} from 32 kW/cm^2, or well above the damage threshold, to 2.2 kW/cm^2, for a laser employing 4-(dicyanomethylene)-2-methyl-6-julolidyl-9-enyl-4H-pyran–doped tris(8-hydroxyquinoline) aluminum if the triplets can be effectively removed from the emissive guest. They hence show that the lasing duration can be extended to nearly 100 μs, ultimately limited by degradation of the lasing medium when a "triplet manager" molecule, 9,10-di(naphtha-2-yl)anthracene, is blended into the gain region of an otherwise conventional distributed feedback OSL. The triplet manager facilitates radiative singlet transfer while suppressing nonradiative triplet transfer to the emitter molecule, thus reducing the triplet-induced losses.

2.6.2 Maximum Pulsewidth for Organic Lasers

The consequence of the absence of observed CW lasing action in organic solid-state lasers is the existence of a finite, maximum pulse width under CW pumping, which was initially observed under flashlamp pumping with μs pump pulses [16].

We assume that triplets are formed through ISC (we use here a constant rate model), at a rate which is much slower than the time needed to reach a steady-state singlet population S_1^{ss}. The dynamics of the singlet states (controlled by a time constant $\sim \tau_F$) is thus supposed to be much faster than the triplet dynamics. Here we also assume that triplet absorption (T–T) results in a loss of laser photons, but not in an appreciable loss of triplet population (see Sect. 2.3.4: it was shown by Yang et al. [74] that 80 % of the high-lying T_n states created by TT absorption in F8BT fall back to the T1 state within 300 fs). With these simplifying assumptions, the triplet population follows:

$$\frac{dT_1}{dt} = k_{ISC} S_1^{ss} - \frac{T_1}{\tau_T}$$

T_1 hence increases in a monoexponential way:

$$T_1(t) = k_{ISC} \tau_T S_1^{ss} \left[1 - e^{-\frac{t}{\tau_T}} \right]$$

until the triplet state population reaches the steady-state value $T_1^{ss} = k_{ISC} \tau_T S_1^{ss}$.

In the case where the CW lasing condition presented above $\sigma_{TT}T_1^{ss} > \sigma_{em}S_1^{ss}$ is not met, then lasing terminates at time t_L given by $t_L = -\tau_T ln\left(1 - \frac{\sigma_{em}}{\sigma_{TT}k_{ISC}\tau_T}\right)$; if t_L is significantly smaller than the triplet lifetime, then this expression becomes:

$$t_L \sim \frac{\sigma_{em}}{\sigma_{TT}k_{ISC}}$$

typical maximum pulse length of an organic laser

As in addition σ_{em} and σ_{TT} have the same order of magnitude, an easy rule-of-thumb is

$$t_L \sim 1/k_{ISC}$$

The pulsewidth is expected to vary from a few ns to a few µs depending essentially on the ISC rate, and has a typical value of ~100 ns.

2.6.3 Temporal Dynamics Simulations

We can go further and solve the coupled rate equations and photon density equation. Here the constant yield approach is used. However numerical simulations revealed only little difference between the two (as discussed earlier, there might be a difference essentially in the saturated regime).

Simulations are shown in Fig. 2.22. The first simulation (1) represents the case of a pure 4-level system, that is $\Phi_T = 0$ (and $\sigma_{TT} = 0$ even though this value does not intervene in the simulation as the triplet state is not populated): the startup behavior is clearly seen here because we choose here a cavity much longer than the active medium, meaning that a photon spends most of its time in air between the two mirrors than in the polymer film, which causes quite a long oscillation buildup time (see Chap. 4 for more details about the so-called VECSOL laser architecture). During this time the singlet density rises to a high value before laser oscillation starts; when the laser intensity reaches the laser saturation intensity I_{sat} (here $I = I_{sat}$ is reached at t = 8.5 ns), the singlet state is rapidly depopulated until it clamps to its steady-state value (here 0.05). The laser enters a CW stable regime from t ~ 20 ns in this particular case.

The simulation (2) has exactly the same parameters except that some triplet filling is added with a fixed quantum yield $\phi_T = 5\%$, but triplets are assumed to be non absorbing, i.e. σ_{TT} is fixed to 0. This is equivalent (only at I = 0 as discussed earlier) to an intersystem crossing rate fixed around 10^7 s^{-1} and a triplet state lifetime of 100 µs, with a radiative singlet lifetime of 5 ns, which are all very typical values. Interestingly, we notice that the triplet population grows up until it quickly becomes the dominant population in the system, meaning that the ground state S_0 is depopulated in favor of the triplet state. Because the pumping intensity is supposed to be constant, the pumping rate (which is proportional to S_0)

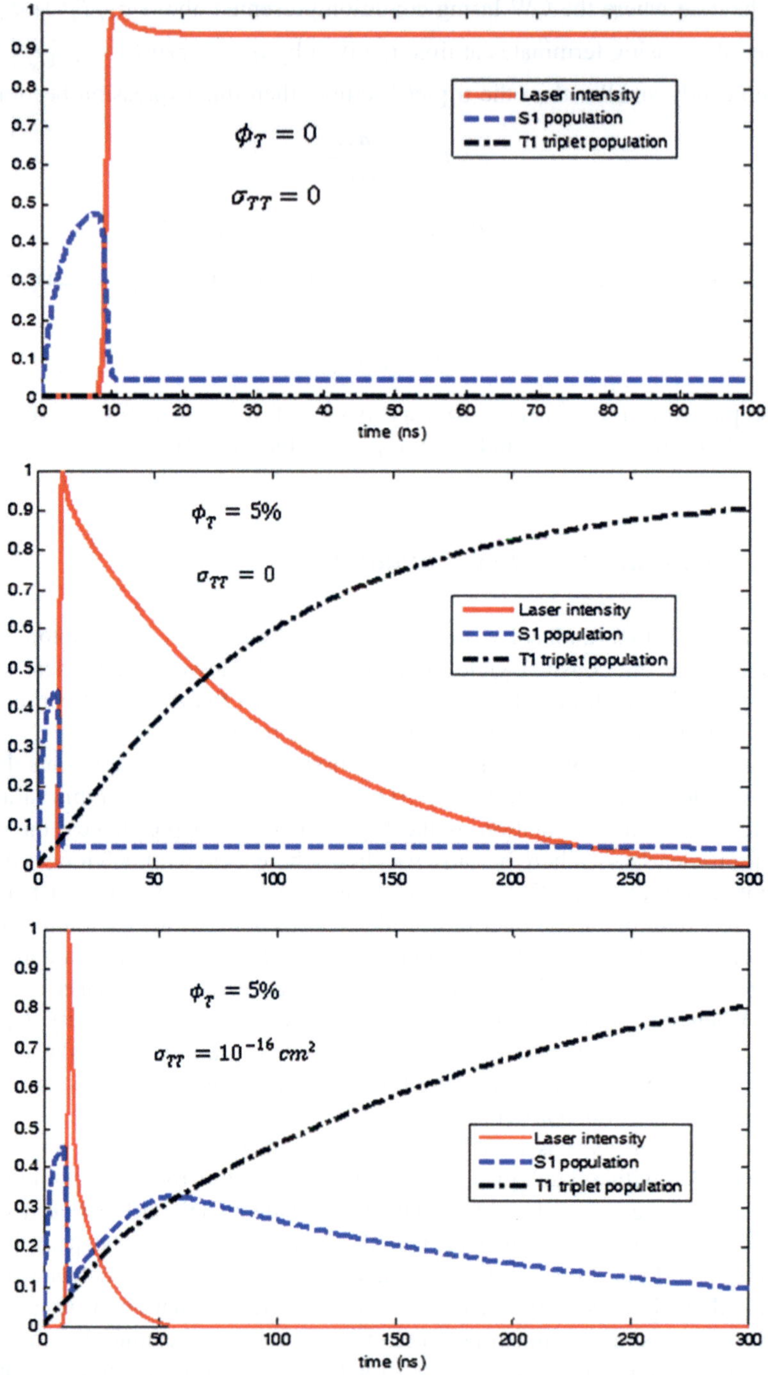

Laser intensity
S1 population
T1 triplet population
$\phi_T = 0$
$\sigma_{TT} = 0$
$\phi_T = 5\%$
$\sigma_{TT} = 0$
$\phi_T = 5\%$
$\sigma_{TT} = 10^{-16}\,cm^2$
time (ns)

◀ **Fig. 2.22** Results of the simulations showing the evolution of the singlet density S1 (= number of molecules in the S1 state/total number of molecules), the triplet density T1 and the normalized laser intensity I versus time in an organic laser pumped with a constant intensity Ip. The specific structure considered here is a VECSOL-type organic laser (see Chap. 4), which allows visualizing the turn-on dynamics with a comparable timescale. The following parameters are taken for the simulation: thickness of the active medium = 10 μm, absorption = 80 %, absorption cross section = 10–16 cm^2; cavity length = 20 mm; outcoupling = 2 % (the reflectivity of the cavity mirrors are respectively taken to 1 and 0.98), passive losses = 2 % per single pass in the gain medium; fluorescence lifetime = 5 ns; emission cross section = 10–16 cm^2; molecular density N = 1019 cm^{-3}. The pump intensity is fixed to the pump saturation intensity $I_p = I_{p,sat} = \frac{hc}{\lambda_p \sigma_{abs} \tau} = 0.5\text{MW}/\text{cm}^2$, that is 20 times above the steady-state threshold Case 1 (*top*): triplet yield and triplet absorption cross sections are fixed to 0 (4-level case); case 2 (*middle*): a fixed triplet yield of 5 % is supposed, but triplets do not absorb laser light; case 3 (*bottom*): a fixed triplet yield of 5 % is considered together with a triplet absorption cross section of 10^{-16}cm^2

decreases, which would tend to decrease the S_1 density. Actually the S_1 density is just reduced below the clamping value which ensures that gain = losses, meaning that the intensity will start to decrease. When this happens, the downward stimulated emission rate from S_1 to S_0 decreases also which tends to stabilize the S_1 density irrespective of a diminution of the pumping rate. This crude model is informative in a sense that even with non-absorbing triplet species or in absence of any triplet quenching phenomenon added to the model, we have a limitation in the pulse duration linked only to the depopulation of the ground state toward the triplet state. Here the laser pulse stops completely after 300 ns.

The simulation (3) at last shows the same system with the same parameters except that some Triplet absorption (TA) is added through a triplet excited state absorption $\sigma_{TT} = 10^{-16}$ cm^2. As expected in this case, the emitted pulse is still shorter, here it stops after 50 ns (but is less than 10 ns FWHM). The singlet density increases when the laser intensity drops significantly, which can be interpreted as a kind a Q-switching provided by the triplet losses. Note however that by using equations above we ignored triplet state depletion by absorption of laser light: although this is certainly true to consider that once in a highly-excited state T_n, the molecules may undergo a rapid internal conversion back to the T_1 state, they may also be removed to dissociative states or undergo internal conversion back to singlet states, among other possibilities. If we suppose that all laser photons absorbed from a T_1 state represent a net loss for the triplet population, then the curves obtained in cases (2) and (3) of Fig. 2.22 would be very similar, meaning that TA would play only a minor role in the CW impediment of organic lasers in such a case.

Recently, the compared influence of Triplet Absorption (TA) versus Triplet-Singlet annihilation (TSA) in the limited pulse width obtainable from an organic laser has become a subject under debate. Giebink and Forrest [67] studied laser dynamics in a DFB laser and related the experiments to TSA only; Lehnhardt [62] showed in a polymer laser made of MEH-PPV with F8BT that TSA and TA were both important, TA becoming notably predominant at high excitation densities.

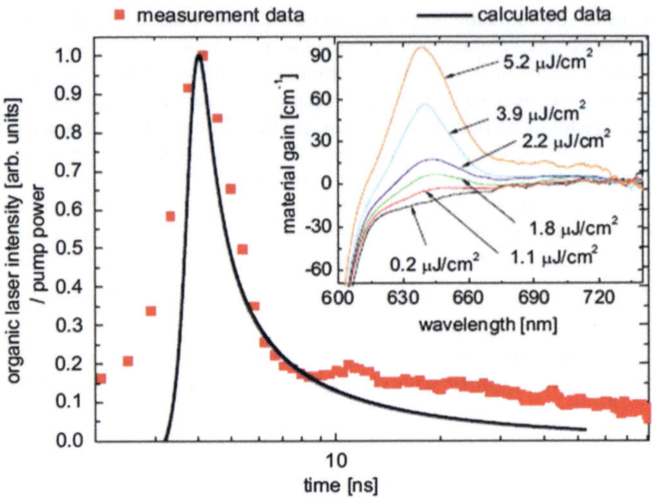

Fig. 2.23 Illustration of the finite pulsewidth obtainable from an organic Solid-state laser. Here the F8BT/MEHPPV laser is pumped by a laser diode with a pulse length of 70 ns; lasing terminates at ~10 ns only. Courtesy of T. Riedl. Reprinted with permission from [62]. Copyright (2010) by the American Physical Society

In such a case, the experimental pulse duration from the laser is only ~3 ns, as shown in Fig. 2.23.

2.6.4 Is It Better to Characterize Organic Lasers with Fluence (J/cm²) or Intensity (W/cm²)?

As organic lasers only emit pulses of light, one may wonder what is the most relevant parameter to measure in a working laser for both the pump and the output beam: is it an energy per pulse (or a related fluence in J/cm^2) or an intensity per pulse (or, in exact radiometric terms, an irradiance) in W/cm^2? In many literature reports, both are used evenly, and this deserves some particular attention.

For someone using the laser, the unit will primarily depend on the targeted application. For nonlinear optics for instance, only the peak power (and hence the intensity) is important, while for bio-applications where a certain amount of light energy is needed to treat a tissue or accumulate data, no matter the time, and of course the energy is more valuable.

However, when characterizing a laser, as stimulated processes (absorption and emission) depend on intensities, it is generally not very relevant to draw physical argumentations based on energies. Let's discuss it in more detail. If we consider a medium pumped by a constant pump intensity I_p from $t > 0$, then the singlet state population will "load", firstly without SE as far as the threshold is not reached yet, with an "effective" time constant given by $\tau_{eff} = \frac{\tau}{1+\sigma_{abs}\tau I_p}$ (obtained from the rate equations with I = 0). The effective loading time can be significantly lower than the radiative lifetime is the pump intensity is higher than the *pump* saturation intensity $1/\sigma_{abs}\tau$ (in photonic units).

If the pump pulse $T_p \ll \tau_{eff}$, then the pump can be seen as a Dirac pulse; in this case, as far as some molecules have been promoted to the S_1 level, they can only emit a laser photon (or a fluorescence photon, or lose their energy by any means), just *once* for one given pump pulse. In this case it is relevant to measure the energy per pulse. The laser is an "energy storing" device, it stores the energy pump during at most a time equal to the radiative lifetime, and delivers part of this stored energy in a single laser pulse.

If $T_p > \tau_{eff}$ then a given emitter can undergo several cycles of absorption/ emission during one single pulse. Then even though a steady state may not be reached within one pulse, the situation is more like a CW laser working intermittently, or an "energy-converting" device which continuously converts the pump light into laser light. In cases where it is easy to measure the output power of the laser, one can compare the number of emitted photons during a pulse to the number of emitters in the active region to count how many times a given emitter has been used during the pulse [98]. In such a case, intensity is the only relevant parameter.

In particular, if one wants to compare the thresholds or the efficiencies of two different lasers, comparing the fluences (in J/cm^2) has no meaning if the pump pulse durations are not identical and both higher than the effective lifetime (e.g. 1 vs. 10 ns), but it will not be problematic to compare the energies with a fs-pump laser and with a ps-pump laser as for the organic medium, these two pump sources are equally seen as ultrashort "Dirac" pulses providing a given amount of energy.

References

1. W. Holzer et al., Spectroscopic and travelling-wave lasing characterisation of Gilch-type and Horner-type MEH-PPV. Synth. Met. **140**(2–3), 155–170 (2004)
2. J.C. Ribierre et al., Amplified spontaneous emission and lasing properties of bisfluorene-cored dendrimers. Appl. Phys. Lett. **91**(081108) (2007)
3. E. Ishow et al., Multicolor emission of small molecule-based amorphous thin films and nanoparticles with a single excitation wavelength. Chem. Mater. **20**(21), 6597–6599 (2008)
4. C.H. Kim et al., Modeling the low-voltage regime of organic diodes: origin of the ideality factor. J. Appl. Phys. **110**(9), 093722 (2011)
5. V. Coropceanu et al., Hole- and electron-vibrational couplings in Oligoacene crystals: intramolecular contributions. Phys. Rev. Lett. **89**(27), 275503 (2002)
6. M. Malagoli, J.L. Bredas, Density functional theory study of the geometric structure and energetics of triphenylamine-based hole-transporting molecules. Chem. Phys. Lett. **327**(1–2), 13–17 (2000)
7. V. Coropceanu et al., Charge transport in organic semiconductors. Chem. Rev. **107**(4), 926 (2007)

8. Y. Shirota, H. Kageyama, Charge carrier transporting molecular materials and their applications in devices. Chem. Rev. **107**(4), 953 (2007)
9. A. Moliton, *Optoelectronics of molecules and polymers* (Springer, New York, 2005)
10. P.M. Borsenberger, L. Pautmeier, H. Bassler, Charge transport in disordered molecular solids. J. Chem. Phys. **94**(8), 5447–5454 (1991)
11. I.I. Fishchuk et al., Nondispersive polaron transport in disordered organic solids. Phys. Rev. B **67**(22), 224303 (2003)
12. L. Li, H. Kosina, Charge transport in organic semiconductor devices, in *Organic Electronics*, ed. by T. Grasser, G. Meller (Springer, Berlin, 2010), p. 301
13. S. Moller, G. Weiser, C. Lapersonne-Meyer, Excitonic photoconductivity of 4BCMU polydiacetylene single crystals. Synth. Met. **116**(1–3), 23–26 (2001)
14. S.F. Alvarado et al., Direct determination of the exciton binding energy of conjugated polymers using a scanning tunneling microscope. Phys. Rev. Lett. **81**(5), 1082–1085 (1998)
15. P.P. Sorokin, R. Lankard, Stimulated emission observed from an organic dye, chloro-aluminium phtalocyanine. IBM J. Res. Develop **10**, 162–163 (1966)
16. F.P. Schafer (ed.), *Dye Lasers*. Topics in Applied Physics, vol. 3, ed. by F.P. Schafer (Springer, Berlin, 1973), p. 285
17. R. Bornemann, U. Lemmer, E. Thiel, Continuous-wave solid-state dye laser. Opt. Lett. **31**(11), 1669 (2006)
18. S. Chénais, S. Forget, Recent advances in solid-state organic lasers. Polym. Int. **61**(3), 390–406 (2012)
19. C.H.J. Wells, Introduction to Molecular Photochemistry. Chapman and Hall Chemistry Textbook Series (Chapman and Hall, New York, 1972)
20. J.M. Holt, *Ultrafast Optical Measurements of Charge Generation and Transfer Mechanisms of Pi-conjugated Polymers for Solar Cell Applications*, University of Utah, 2009
21. P. Chaquin, F. Fuster, Orbimol Laboratoire de Chimie Théorique, (UPMC Univ Paris 6, UMR CNRS 7616, Paris, 2012), http://www.lct.jussieu.fr/pagesperso/orbimol/
22. H.A.M. van Mullekom et al., Developments in the chemistry and band gap engineering of donor-acceptor substituted conjugated polymers. Mater. Sci. Eng. R Reports **32**(1), 1–40 (2001)
23. J. Roncali, Molecular engineering of the band gap of π-conjugated systems: facing technological applications. Macromol. Rapid Commun. **28**(17), 1761–1775 (2007)
24. C.A. Coulson, Excited electronic levels in conjugated molecules: I. Long wavelength ultra-violet absorption of naphthalene, anthracene and homologs. Proc. Phys. Soc. **60**(3), 257 (1948)
25. T.H. Fay, S.D. Graham, Coupled spring equations. Int. J. Math. Educ. Sci. Technol. **34**(1), 65–79 (2003)
26. A.C. Tropper et al., Vertical-external-cavity semiconductor lasers. J. Phys. D Appl. Phys. **37**(9), R75 (2004)
27. S. Jasprit, *Electronic and Optoelectronic Properties of Semiconductor Structures* (Cambridge University Press, Cambridge, 2007), p. 560
28. G.G. Malliaras et al., Nondispersive electron transport in Alq[sub 3]. Appl. Phys. Lett. **79**(16), 2582 (2001)
29. T. Susdorf et al., Photophysical characterisation of some dipyrromethene dyes in ethyl acetate and covalently bound to poly(methyl methacrylate). Chem. Phys. **312**(1–3), 151–158 (2005)
30. I. Gozhyk et al., Polarization properties of solid-state organic lasers. Phys. Rev. A **86**(4), 043817 (2012)
31. I. Gozhyk et al., Towards polarization controlled organic micro-lasers. in *Photonics West* (SPIE, San Francisco, 2012)
32. B. Valeur, *Molecular Fluorescence* (Wiley-VCH, Weinheim, 2001)
33. H.-W. Lin et al., Tuning stimulated emission of organic thin films by molecular reorientation. Appl. Phys. Lett. **87**(7), 071910–071913 (2005)
34. F.J. Duarte, *Tunable Laser Applications*, 2nd edn. (CRC Press, New York, 2009)

35. M. Goossens et al., Subpicosecond pulses from a gain-switched polymer distributed feedback laser. Appl. Phys. Lett. **85**(1), 31 (2004)
36. P. Atkins, J.D. Paula, *Physical Chemistry* (Oxford University Press, New York, 2006)
37. M. Fox, *Optical Properties of Solids*. (Oxford Master Series in Condensed Matter Physics) (Oxford University Press, New York, 2002)
38. S. Forget et al., Red-emitting fluorescent organic light emitting diodes with low sensitivity to self-quenching. J. Appl. Phys. **108**(064509) (2010)
39. S.J. Strickler, R.A. Berg, Relationship between absorption intensity and fluorescence lifetime of molecules. J. Chem. Phys. **37**(4), 814–822 (1962)
40. W. Holzer, A. Penzkofer, T. Tsuboi, Absorption and emission spectroscopic characterization of Ir(ppy)(3). Chem. Phys. **308**(1–2), 93–102 (2005)
41. M. Reufer, J.M. Lupton, U. Scherf, Stimulated emission depletion of triplet excitons in a phosphorescent organic laser. Appl. Phys. Lett. **89**(14), 141111–141113 (2006)
42. S.W. Hell, J. Wichmann, Breaking the diffraction resolution limit by stimulated emission: stimulated-emission-depletion fluorescence microscopy. Opt. Lett. **19**(11), 780–782 (1994)
43. M.D. McGehee et al., Semiconducting polymer distributed feedback lasers. Appl. Phys. Lett. **72**(13), 1536–1538 (1998)
44. S. Chandra et al., Tunable ultraviolet laser source based on solid-state dye laser technology and CsLiB6O10 harmonic generation. Opt. Lett. **22**(4), 209 (1997)
45. S. Forget et al., Tunable ultraviolet vertically-emitting organic laser. Appl. Phys. Lett. **98**(13), 131102 (2011)
46. X.H. Yang et al., Highly efficient polymeric electrophosphorescent diodes. Adv. Mater. **18**(7), 948 (2006)
47. Y.F. Pedash et al., Spin-orbit coupling and luminescence characteristics of conjugated organic molecules. I. Polyacenes. J. Mol. Struct. (Thoechem) **585**(1), 49–59 (2002)
48. R.F. Kubin, A.N. Fletcher, The effect of oxygen on the fluorescence quantum yields of some coumarin dyes in ethanol. Chem. Phys. Lett. **99**(1), 49–52 (1983)
49. G. Tsiminis et al., A two-photon pumped polyfluorene laser. Appl. Phys. Lett. **94**(25), 253304 (2009)
50. Y. Mo et al., Ultraviolet-emitting conjugated polymer poly(9,9[prime or minute]-alkyl-3,6-silafluorene) with a wide band gap of 4.0 eV. Chem. Commun. **39**, 4925 (2005)
51. N. Johansson et al., Solid-state amplified spontaneous emission in some spiro-type molecules: a new concept for the design of solid-state lasing molecules. Adv. Mater. **10**(14), 1136 (1998)
52. T. Spehr et al., Organic solid-state ultraviolet-laser based on spiro-terphenyl. Appl. Phys. Lett. **87**(16), 161103 (2005)
53. J.V. Caspar, T.J. Meyer, Application of the energy gap law to nonradiative, excited-state decay. J. Phys. Chem. **87**(6), 952–957 (1983)
54. P. Del Carro et al., Near-infrared imprinted distributed feedback lasers. Appl. Phys. Lett. **89**(20), 201105 (2006)
55. S. Yuyama et al., Solid state organic laser emission at 970 nm from dye-doped fluorinated-polyimide planar waveguides. Appl. Phys. Lett. **93**(2), 023306 (2008)
56. M. Casalboni et al., 1.3 mu m light amplification in dye-doped hybrid sol-gel channel waveguides. Appl. Phys. Lett. **83**(3), 416 (2003)
57. C. Winder, N.S. Sariciftci, Low bandgap polymers for photon harvesting in bulk heterojunction solar cells. J. Mater. Chem. **14**(7), 1077–1086 (2004)
58. R.E. Peierls, *Quantum Theory of Solids* (Oxford University Press, London, 1956)
59. H. Föll, http://www.tf.uni-kiel.de/matwis/amat/semi_en/kap_a/advanced/ta_4_1.html
60. S.A. Jenekhe, A class of narrow-band-gap semiconducting polymers. Nature **322**(6077), 345–347 (1986)
61. M.A. Baldo et al., Highly efficient phosphorescent emission from organic electroluminescent devices. Nature **395**(6698), 151–154 (1998)

62. M. Lehnhardt et al., Impact of triplet absorption and triplet-singlet annihilation on the dynamics of optically pumped organic solid-state lasers. Phys. Rev. B **81**(16), 165206 (2010)
63. Y. Zhang, S.R. Forrest, Existence of continuous-wave threshold for organic semiconductor lasers. Phys. Rev. B **84**(24), 241301 (2011)
64. S. Schols et al., Triplet excitation scavenging in films of conjugated polymers. Chem. Phys. Chem. **10**(7), 1071–1076 (2009)
65. S. Kéna-Cohen et al., Plasmonic sinks for the selective removal of long-lived states. ACS Nano **5**(12), 9958–9965 (2011)
66. M.A. Baldo, R.J. Holmes, S.R. Forrest, Prospects for electrically pumped organic lasers. Phys. Rev. B **66**(3), 035321 (2002)
67. N.C. Giebink, S.R. Forrest, Temporal response of optically pumped organic semiconductor lasers and its implication for reaching threshold under electrical excitation. Phys. Rev. B **79**(073302) (2009)
68. N.C. Giebink, Y. Sun, S.R. Forrest, Transient analysis of triplet exciton dynamics in amorphous organic semiconductor thin films. Org. Electron. **7**(5), 375–386 (2006)
69. A. Kohler, H. Bassler, Triplet states in organic semiconductors. Mater. Sci. Eng. R Reports **66**(4–6), 71–109 (2009)
70. S.P. McGlynn, T. Azumi, M. Kinoshita, Molecular Spectroscopy of the Triplet State, ed.(P.-H. International, Hemel Hempstead, 1969). ISBN: 0135996627
71. A. Köhler, D. Beljonne, The singlet–triplet exchange energy in conjugated polymers. Adv. Funct. Mater. **14**(1), 11–18 (2004)
72. J. Michl, E.W. Thulstrup, Why is azulene blue and anthracene white? A simple mo picture. Tetrahedron **32**(2), 205–209 (1976)
73. C. Cohen-Tannoudji, B. Diu, F. Laloe, *Quantum mechanics/Claude Cohen-Tannoudji, Bernard Diu, Franck Laloe; translated from the French by Susan Reid Hemley, Nicole Ostrowsky, Dan Ostrowsky* (Wiley, New York, 1977)
74. X. Yang et al., Saturation, relaxation, and dissociation of excited triplet excitons in conjugated polymers. Adv. Mater. **21**(8), 916–919 (2009)
75. C. Adachi et al., Nearly 100% internal phosphorescence efficiency in an organic light-emitting device. J. Appl. Phys. **90**(10), 5048–5051 (2001)
76. S. Schols, *Device Architecture and Materials for Organic Light-Emitting Devices*, 1st edn. (Springer, Berlin, 2011). ISBN: 9789400716070
77. J. Lakowicz, *Principles of Fluorescence Spectroscopy*, 3rd edn. (Springer, New York, 2006)
78. R.M. Clegg, Förster resonance energy transfer—FRET what is it, why do it, and how it's done, in *Laboratory Techniques in Biochemistry and Molecular Biology*, vol. 33, ed. by T.W.J. Gadella, (Elsevier, Amsterdam, 2009), pp. 1–57
79. D.F. Evans, 257. Perturbation of singlet-triplet transitions of aromatic molecules by oxygen under pressure. J. Chem. Soc. (Resumed), **1957**, 1351–1357 (1957)
80. M. Lebental et al., Diffusion of triplet excitons in an operational organic light-emitting diode. Phys. Rev. B **79**(165318) (2009)
81. R.R. Lunt, et al., Exciton diffusion lengths of organic semiconductor thin films measured by spectrally resolved photoluminescence quenching. J. Appl. Phys. **105**(053711) (2009)
82. G. Weber, Dependence of the polarization of the fluorescence on the concentration. Trans. Faraday Soc. **50**, 552–555 (1954)
83. S.Y. Arzhantsev et al., On the singlet–singlet annihilation of the excited states of Rhodamine 3B in a polymer film. Laser Phys. **9**(2), 466–469 (1999)
84. C. Gärtner, *Organic Laser Diodes: Modelling and Simulation* (Universitätsverlag Karlsruhe, Karlsruhe, 2009)
85. M.A. Baldo et al., Excitonic singlet-triplet ratio in a semiconducting organic thin film. Phys. Rev. B **60**(20), 14422–14428 (1999)
86. E.J.W. List et al., Direct evidence for singlet-triplet exciton annihilation in Ï€-conjugated polymers. Phys. Rev. B **66**(23), 235203 (2002)

87. M.A. Baldo, C. Adachi, S.R. Forrest, Transient analysis of organic electrophosphorescence. II. Transient analysis of triplet–triplet annihilation. Phys. Rev. B. **62**(16), 10967–10977 (2000)
88. A. Siegman, *Lasers* (University Science Books, Mill Valey, 1986)
89. M. Koschorreck et al., Dynamics of a high-Q vertical-cavity organic laser. Appl. Phys. Lett. **87**(181108) (2005)
90. L.A. Coldren, S.W. Corzine, *Diode Lasers and Photonic Integrated Circuits* (Wiley-VCH, New York, 1997)
91. C. Delsart, *Lasers & Optique Non Linéaire* (Ellipses, Paris, 2008), p. 426
92. C. Gartner et al., The influence of annihilation processes on the threshold current density of organic laser diodes. J. Appl. Phys. **101**(2), 023107 (2007)
93. G. Lanzani et al., Triplet-exciton generation mechanism in a new soluble (Red-Phase) Polydiacetylene. Phys. Rev. Lett. **87**(18), 187402 (2001)
94. J. Widengren, R. Rigler, Ü. Mets, Triplet-state monitoring by fluorescence correlation spectroscopy. J. Fluoresc. **4**(3), 255–258 (1994)
95. A. Penzkofer, W. Falkenstein, Theoretical investigation of amplified spontaneous emission with picosecond light pulses in dye solutions. Opt. Quant. Electron. **10**(5), 399–423 (1978)
96. C. Gartner et al., The influence of annihilation processes on the threshold current density of organic laser diodes. J. Appl. Phys. **101**(2), 023107–023109 (2007)
97. C. Zenz et al., Ultrafast photogeneration mechanisms of triplet states in para-hexaphenyl. Phys. Rev. B **59**(22), 14336–14341 (1999)
98. H. Rabbani-Haghighi et al., Highly efficient, diffraction-limited laser emission from a vertical external-cavity surface-emitting organic laser. Opt. Lett. **35**(12), 1968–1970 (2010)
99. D. Cahen, A. Kahn, E. Umbach, Energetics of molecular interfaces. Mater. Today **8**(7), 32–41 (2005)
100. H.A.M. van Mullekom et al., Developments in the chemistry and band gap engineering of donor-acceptor substituted conjugated polymers. Mater. Sci. Eng. R Reports **32**(1), 1 (2001)
101. M.P. Lettinga, H. Zuilhof, M.A.M.J. van Zandvoort, Phosphorescence and fluorescence characterization of fluorescein derivatives immobilized in various polymer matrices. Phys. Chem. Chem. Phys. **2**(16), 3697–3707 (2000)
102. T.G. Pavlopoulos et al., Laser action from syn-(methyl, methyl) bimane. J. Appl. Phys. **60**(11), 4028–4030 (1986)
103. A. Costela et al., Polymeric matrices for lasing dyes: recent developments. Laser Chem. **18**(1–2), 63–84 (1998)
104. J. Yu et al., Singlet-triplet and triplet–triplet interactions in conjugated polymer single molecules. J. Phys. Chem. B **109**(20), 10025–10034 (2005)
105. M.A. Stevens et al., Exciton dissociation mechanisms in the polymeric semiconductors poly(9,9-dioctylfluorene) and poly(9,9-dioctylfluorene-co-benzothiadiazole). Phys. Rev. B **63**(16), 165213 (2001)
106. E.J.W. List et al., Direct evidence for singlet-triplet exciton annihilation in pi-conjugated polymers. Phys. Rev. B **66**(235203) (2002)
107. G.D. Hale, S.J. Oldenburg, N.J. Halas, Observation of triplet exciton dynamics in conjugated polymer films using two-photon photoelectron spectroscopy. Phys. Rev. B **55**(24), R16069–R16071 (1997)
108. D. Hertel, K. Meerholz, Triplet-polaron quenching in conjugated polymers. J. Phys. Chem. B **111**(42), 12075–12080 (2007)

Chapter 3
Organic Materials for Solid-State Lasers

Abstract In this chapter we describe organic materials used in laser systems. We define a classification for organic molecules, and give some details about the chemical structure and the optical properties of some of the main families widely used as gain materials or host matrices in organic lasers. The technological processes allowing organic molecule deposition in various forms are then reviewed, as well as the main experimental techniques used to measure crucial parameters such as optical gain.

3.1 Molecular Materials Suitable for Laser Action

3.1.1 Overview of the Different Families of Materials Used in Lasers

Dye molecules dissolved in liquid solvents have been for decades the winning scheme for organic lasers [1]. The first attempts to make solid-state organic lasers naturally consisted in using the same compounds. A "**dye**" can be defined as a π-conjugated molecule with a high quantum yield of fluorescence, it can be either neutral or ionic: typical examples are xanthenes (rhodamine and fluorescein families), coumarines, oxazines, or pyrromethenes [2]. Even though a given dye molecule can be a very good light emitter in a dilute form, it will be less efficient or even non-emissive at all at high concentrations [2] because of intermolecular interactions arising between nearby molecules discussed in Sect. 2.4, referred to as "concentration quenching". In the solid state, a straightforward strategy to avoid quenching is to incorporate dye molecules as dopants into solid matrices. Solid hosts can be either polymers—for instance poly(methyl methacrylate) (PMMA) and its derivatives [3]—, glasses, hybrid organic–inorganic materials prepared by sol-gel techniques [4–7], etc.

S. Forget and S. Chénais, *Organic Solid-State Lasers*,
Springer Series in Optical Sciences 175, DOI: 10.1007/978-3-642-36705-2_3,

Later, organic semiconductors (OSC) have been identified as suitable for lasing. The term refers to their electrical semiconducting properties, although the physics of electrical conduction in organics differs notably from that of inorganic semiconductors (see Sect. 2.1.2 and [8]). Based on structural considerations, one can distinguish three categories of organic semiconductors: organic crystals, small molecules and polymers.

Organic single crystals of anthracene or pentacene [9, 10] resemble inorganic crystals, and their transport properties can be defined in terms of bands. In practice, the high voltage needed to get light from those crystals limited their practical interest and their difficult growth limited their lasing applications as well. In addition, high molecular-crystal packing provides high mobilities which is generally in contradiction with a high quantum yield of fluorescence [11]. Single crystals are however being considered for light-emitting transistors [12] for example.

Small-molecular organic semiconductors are at the origin of the astonishing development of commercial thermally-evaporated OLEDs [13, 14]. In such OSCs, which are mostly amorphous, transport occurs by hopping between localized sites, with a probability that critically depends on the orbital intermolecular overlap. Any neat film of a π-conjugated compound will then have semiconducting properties to some extent. However the tendency to intermolecular π–π stacking and dipole–dipole interactions between units leads *a priori* to severe quenching and is detrimental for lasing: many organic semiconductors turn out to be either not emissive or may emit fluorescence but are not suitable for lasing. Many efforts in molecular design have hence been directed towards improving the luminescence properties of organic semiconductor films. This can be done by designing molecular geometries where fluorophore units are kept apart and/or π–π stacking highly reduced [15–17]. Dendrimers for instance are good candidates for organic lasing: the dendrimer structure [18, 19] consists of a chromophore (for example pyrene [20]) located in the core of a highly branched structure ended with surface groups. The core defines the optical properties such as emission wavelength, whereas the surface groups confer solubility and act as spacers to limit π–π stacking. Similarly, the spiro-linkage used to couple two oligomers in spiro-compounds [21–23] enables defining a twisted geometry [24] in which the optical properties of the individual moieties are conserved. However the easiest way to get rid of concentration quenching is still to physically separate the emissive units; it is possible however to give to the host an active role with host/guest systems. The archetypal example is Alq_3/DCM [25, 26]. In this case, the pump light is absorbed by the higher energy gap host material (Alq_3, Fig. 3.1a), which very efficiently transfers its energy to the lower energy gap guest (DCM, Fig. 3.1d) through a Förster mechanism [27]. Compared to DCM doped into a passive matrix (like PMMA), the host/guest structure enables a much higher absorption of the pump; the obtained material is furthermore able to transport charges through Alq_3. At last, the Stokes shift between absorption and emission wavelength is increased, which consequently reduces self-absorption (i.e. re-absorption of the emitted laser light).

The other major category of semiconducting organic molecules is formed by **conjugated polymers** [28]. Conjugated polymers can exhibit high photoluminescence

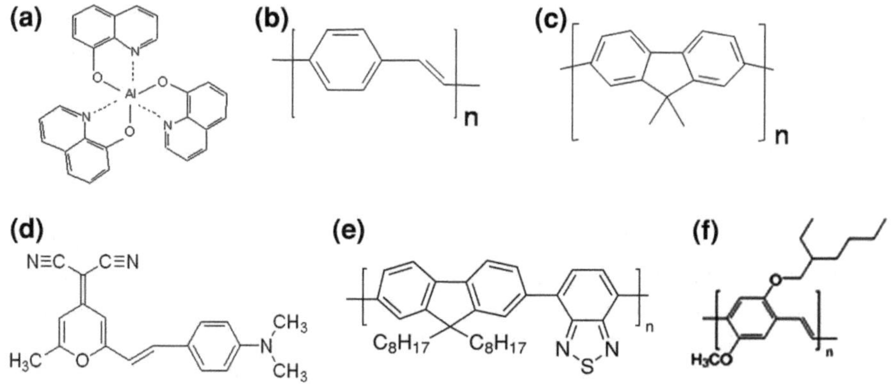

Fig. 3.1 Chemical structure of some typical organic media used in organic lasers: **a** aluminium tris(quinolate) or Alq_3, **b** generic polyphenylene vinylene or PPV, **c** generic polyfluorene, **d** DCM (4-(Dicyanomethylene)-2-methyl-6-(4-dimethylaminostyryl)-4H-pyran, **e** F8BT (poly(9,9'-dioctylfluorene-co-benzothiadiazole)), **f** MEH-PPV (poly(2-méthoxy-5-(2-éthyl-hexyloxy)-1,4-phenylene-vinylene))

yields even in the solid-state. Lasing has been demonstrated in a large number of such macromolecules, especially in poly(phenylene vinylene) (PPV) [29–31] and polyfluorenes [32, 33] (Fig. 3.1b and c). Host/guest systems composed of two polymers have also been recently investigated for laser action, based on MEH-PPV doped into F8BT [34] (Fig. 3.1e and f).

Recently, new materials emerged as good candidates for organic lasing: this is the case of **liquid crystals** [35], whose self-organized structure forms photonic bandgaps that can be exploited for lasing; the gain unit can be a classical dye incorporated in the structure or it might be the liquid crystal itself [36].

A more **exotic lasing material** was newly reported by M. Gather et al., who realized a single-cell "living" laser [37], in which the gain medium is Green Fluorescent Protein, a well-known imaging tracer in biophotonics. The cavity is a high Q microresonator made by two DBR gratings in a vertical microcavity configuration. In addition of its future possible implications in bio-imaging, it presents some new and intriguing aspects, such as the fact that the cell remains alive during laser operation, and the possibility to create a biological self-generating and self-healing laser medium [38].

3.1.2 Towards a Classification of Organic Laser Materials?

Is there a way to classify these different materials (dyes in inert matrices, host/guest systems, conjugated polymers, small molecules, dendrimers, and so forth) in classes that would be relevant for the laser physicist?

A first distinction can be drawn on the basis of **processing techniques**, leading to the differentiation between bulk gain media and thin films. *Bulk gain media* are millimetric to centimetric rods that are intended to be set in macroscopic laser cavities after being polished to optical quality [39]; the rods are fabricated by inserting an organic emitter into a monomer (such as MMA) before polymerization [3, 39] or into a glass by a sol-gel process [40]. In contrast, *thin films* are somewhat easier to implement, either through thermal evaporation or solution processing. Small molecules (including dyes) can be evaporated provided that they are neutral, while polymers are too heavy to do so. Solution processing includes techniques such as spin-coating [41], dip-coating, doctor-blading [42], or ink-jet printing [43]. Many conjugated polymers are solution-processable, as well as small-molecular compounds, provided that they are soluble. Some unsoluble polymers may be made solution processable via e.g. side chain functionalization or main chain copolymerization with soluble segments. Non-conjugated polymers such as PMMA are also readily solution-processable, making dye-doped polymeric lasers also suitable for thin-film devices.

Beyond this "technological" classification, a long-admitted distinction is traditionally made between "dyes" and "organic semiconductors", a distinction that we have discussed in Sect. 2.1.2.3. The problem arising with this distinction is, although it is practical and easy to use with most materials (Pyrromethene small molecules clearly fall into the "dye" category while Polyfluorene as a conjugated polymer is clearly an OSC), some of them are both dyes and OSCs [15, 16, 44].

However, a distinction can still be made at the macroscopic scale of the gain medium rather than at the molecular scale, based on the **importance of intermolecular interactions**, and their influence on laser performance. In a first category, including neat films and host/guest systems (which can be identified to OSCs), exciton–exciton annihilation, exciton diffusion, photogeneration of charges and related phenomena [45–48] (see Sect. 2.4) quench the excited state population, and make the photophysical description of the laser medium different from a simple 4-level picture (cf. Sect. 2.5). This has been shown extensively in conjugated polymers [45, 49], and this is also true in guest/host systems: for instance in the Alq_3: DCM blend, the fast laser dynamics are affected by the Förster transfer dynamics [50] as well as by quenching of the guest (DCM) singlets on photogenerated Alq_3 triplets [51]. In contrast, when an emitter is dissolved in a passive matrix, intermolecular effects are lower and laser physics are somewhat simpler. In that sense, a physically-relevant distinction can be made between "dispersed" gain elements, in which intermolecular interactions, at least the short-range interactions (cf. Sect. 2.4.1) are suppressed, and "neat solids", where all intermolecular phenomena may happen, and which all exhibit, to some extent, organic semiconducting properties.

In the following we keep the traditional distinction, we call "dyes" the compounds which are mostly used in a dispersed form, mostly in nonconductive polymer hosts, and OSCs the remaining.

3.1.3 Small-Molecular Dyes

An organic dye, in the usual and historical sense of the word, is a substance able to impart color to another one. For optical applications, this definition is sharpened and organic dyes are compounds having an absorption band in the spectral region from the UV to the near infrared. All these molecules contain multiple conjugated double bonds, which strongly define their chemical reactivity and their spectral properties. For example, molecules with only single bonds generally absorb light at wavelengths below 150 nm, an energy high enough to induce photodecomposition of the compound which makes such systems useless for lasing applications.

The huge panel of organic molecules able to act as laser dyes is one of the major explanation for the popularity of dye lasers. It allows spectral emission covering the whole visible spectrum, and even near UV and near IR. On commercial websites, more than 500 different dyes can be found, with various emission and absorption wavelengths. Additionally, each dye is able to emit typically over a few tens of nanometer, providing tunability. The exact value of the emitted wavelength at the top of the broad emission curve depends on the chemical structure of the organic material of course, but also on the interaction with the surrounding molecules (solvent in the case of liquid dye laser, polymers for solid-state devices).

Laser dyes are large molecules with an extended system of conjugated bonds (generally including numerous ring structures) leading to complex spectra for both absorption and emission. The usual classification is based on their chemical structure, and also corresponds to a given wavelength coverage. The main families are coumarins, xanthenes, oxazines and pyrromethenes, and the spectral emission is depicted on Fig. 3.2 with some typical structures.

For effective performance, dye molecules should have a strong absorption at pump wavelength and minimal absorption at lasing wavelength to minimize reabsorption losses for the emitted photons. A high quantum yield (as close as 1 as

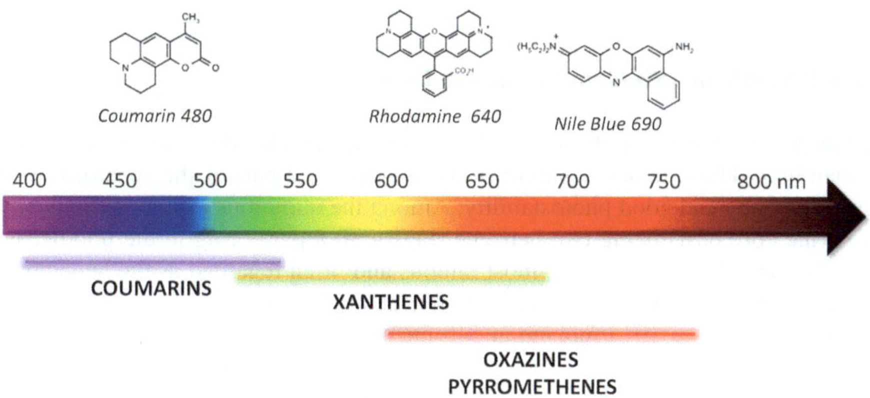

Fig. 3.2 Tunability range for some organic dyes. The wavelength chart indicates the emission of compounds depicted above

possible) is of course highly important for efficient laser emission, together with a low intersystem crossing rate to minimize losses induced by triplet–triplet absorption. A large Stokes shift is desirable to avoid reabsorption losses. A good thermo- and photochemical stability is also crucial, as well as a low tendency of the dye molecule to self-aggregate (this point is essential if high concentration of the dye has to be used).

Laser dyes discovery was often made by trial and errors, but some general rules to obtain efficient laser dyes can be followed. It is for example generally beneficial to design rigid molecules or to work on the planarity of the compounds to enhance fluorescence or to reduce quenching effects.

Historically, the xanthene family has been broadly used, especially the very efficient rhodamine dyes (rhodamine 6G, rhodamine 640...). Over the last decades, dyes have been more specifically developed for laser operation, such as pyrromethenes (or BODIPY dyes) with enhanced performance and photostability. Both rhodamines and pyrromethenes emit in the yellow–red part of the spectrum, making this spectral area the better choice for efficient dye lasers.

We will now rapidly review some of the well-known commercially available laser dyes.

3.1.3.1 Coumarin Laser Dyes

Coumarin dyes form an important family as they can emit light from the violet to the green part of the spectrum. A generic coumarin is the association of a benzene ring with a pyrone ring, with various substituents attached on this basic core (the basic coumarin molecule is not fluorescent) as described in Fig. 3.3. Spectroscopic properties of coumarin dyes are strongly dependant on the nature of these substituents. An example is given in Fig. 3.3 with three examples (Coumarin 480, 500 and 560). Coumarin are relatively efficient dye laser materials, however their photostability is usually weak.

3.1.3.2 Xanthene (Rhodamines) Laser Dyes

Xanthenes are the basis of a class of dyes which includes fluorescein, eosins and rhodamines. These molecules emit in the green to red part of the spectrum, with high efficiency and good photostability. Among the numerous available rhodamine dyes (the core basis being two benzene rings with a pyran ring in the middle, see Fig. 3.4), Rhodamine 6G is the most famous and often regarded as the benchmark for all other laser dyes. Its singlet–singlet absorption is very strong (extinction coefficient of 1.1×10^5 l/mol.cm at 540 nm), but Triplet–triplet absorption (TA), which is one of the major issues of organic lasers (see Sect. 2.3.4), cannot be neglected, as it is typically only 10 times lower [52]. An interesting way of

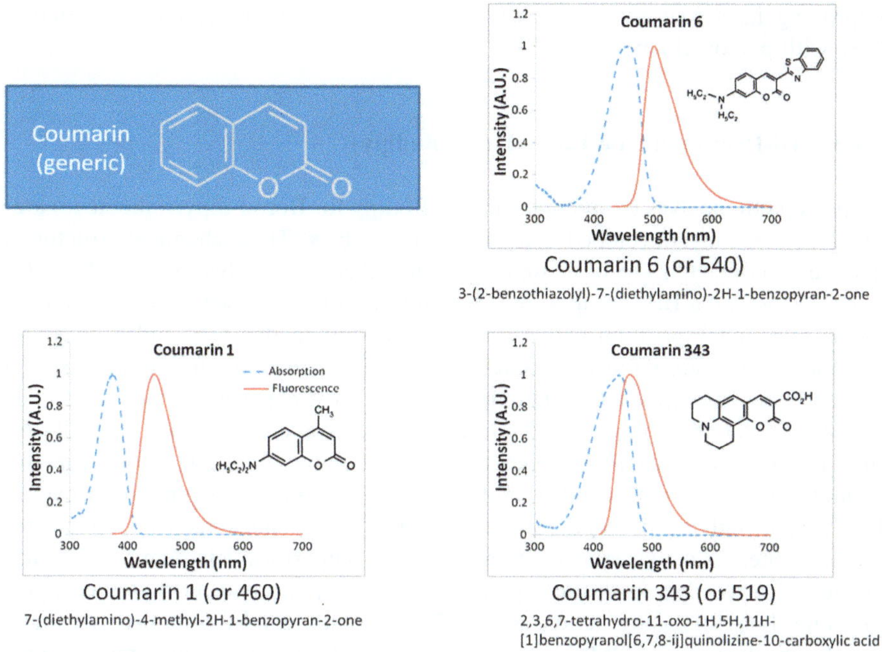

Fig. 3.3 Example of coumarin dyes (in brackets an alternative name sometime used for commercial products) and their absorption (*dotted line*) and fluorescence (*full line*) spectra

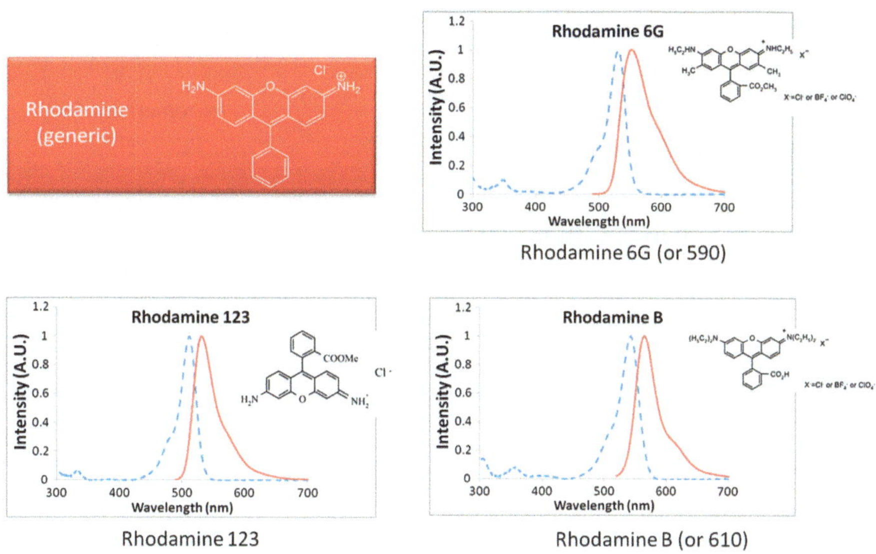

Fig. 3.4 Example of Rhodamine dyes (in brackets an alternative name sometime used for commercial products) and their absorption (*dotted line*) and fluorescence (*full line*) spectra

improving laser dye efficiency is to decrease this T–T absorption, as explained later with pyrromethenes.

3.1.3.3 Di-Pyrromethene Laser Dyes (Bodipy)

Di-pyrromethene dyes (sometimes named Bodipy for Boron-dipyrromethene) are a relatively recently developed category of laser dyes. Their chemical structure is depicted on Fig. 3.5. The basic unit is planar and consists of two pyrroles linked by a methylene and a BF_2 group to provide high rigidity. It presents methyl groups at the 1, 3, 5 and 7 position and various substituents at the 2, 6, and 8 ones. The presence of the aromatic heterocycles is usually a promise for reduced triplet-related losses, as explained by Drexhage [53]. This is experimentally verified with a triplet–triplet absorption over the laser emission wavelength range typically five time less intense in pyrromethenes than in rhodamines [52, 54]. Moreover, they exhibit a good thermal stability and solubility in organic solvents (and even in PMMA), a high fluorescence quantum yield (>0.9) and a reduced intersystem crossing rate. As a consequence, they generally outperform Rhodamine 6G (taken as the reference dye for lasing operation). Probably the two most studied pyrromethene dyes for lasing action are 1,3,5,7,8, pentamethyl-2,6 diethylpyrromethene-difluoroborate (or PM567, named after its fluorescence maximum around 567 nm), where the two lateral R1 groups are C_2H_5, and 1,3,5,7,8, pentamethyl-2,6 di-t-butylpyrromethene-difluoroborate (or PM597), where the two lateral R1

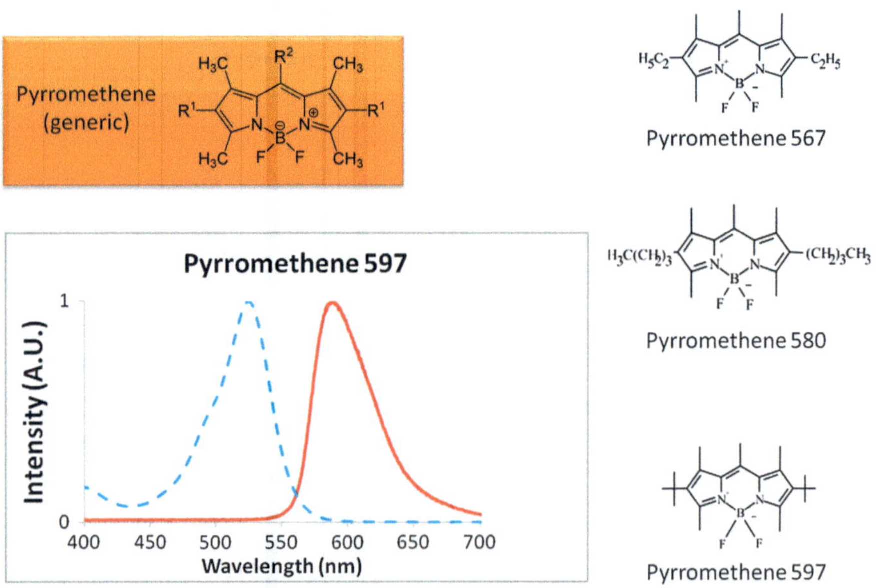

Fig. 3.5 Example of pyrromethene dyes and their absorption (*dotted line*) and fluorescence (*full line*) spectra

groups are tert-butyles. PM597 has a lower quantum yield than PM567, presumably because the steric hindrance induced by the bulky tert-butyl groups that make the molecule non-planar in the exited state [55]. However, on the other hand, this results in an increase Stokes shift favorable for laser action.

Numerous studies have been made to quantify the influence of the substitution at the 8 position (symmetric to the BF2 group) of different chemical compounds such as alkyl, aryl [56] or more complex groups [55, 57]. Pyrromethenes photostability is known to be relatively weak in solution. Indeed, the triplet state of the excited molecule can interact with singlet oxygen at the 7, 8 double bond of the central ring. It produces an unstable compound and eventually the destruction of the dye structure. The influence of oxygen on the degradation is evidenced by the increase of the compound lifetime when oxygen quenchers are added or in deoxygenated solutions. Substitution at the 8 position of different groups also influences the photostability without modifying the lasing properties [58]. Interestingly, the lifetime is higher when the pyrromethene dye is incorporated into a solid-state matrix such as PMMA, probably because the quantity of dissolved oxygen is lower.

3.1.4 Host Materials for Small-Molecular Dyes

Small molecular dyes described above cannot generally emit in neat films, because of concentration quenching issues mentioned in Sect. 2.4. As a consequence, organic molecules must be spatially separated, which can be easily done by incorporating them in passive host matrices.

3.1.4.1 Transparent Nonconductive Polymers as Hosts

The most simple example of an inert solid matrix is a nonconductive highly-transparent polymer. This solid polymer has to fulfill some requirements to be considered as an effective host: it has to be chemically compatible with dyes, transparent and highly homogeneous from an optical point of view, easy to process, and optically and mechanically stable in time. Finally, as they will be subjected to intense photopumping, their damage threshold and thermal properties (lensing effect for example due to potentially high dn/dT [59]) have to be carefully studied. In particular, high purity is needed as every impurity or defect can be seen as an absorbing center able to induce damage in the material. Intense progress has been made in the chemical preparation of the polymer (distillation, sonication, filtration) to improve the resistance of the materials to optical damage. The group of Costela, Garcia-Moreno and Sastre in Spain for example has done an impressive work to enlighten the parameters able to improve the photostability of polymeric matrices. They show that variations around the archetypal and widely used PMMA (using copolymers of MMA) allows some improvement on the lifetime of organic lasers based on rhodamine or pyrromethene dyes, and that for each dye a specific polymer is needed as a matrix to optimize the lasing performances, and especially the operational lifetime [3]. As an example, Pyrromethene 567 dissolved in P(MMA-TFMA 7:3) as a lifetime 6 times shorter than the same dye in a P(MMA-PETA 95:5) matrix [60].

3.1.4.2 Organic–Inorganic Hosts

Organic dyes can also be dispersed in silicate-based organic–inorganic hybrid polymers, to take advantage of the improved thermal and mechanical properties of those compounds. These hybrids can be fabricated by a sol-gel technique and can lead to very high lasing efficiencies when pyrromethenes or perylene dyes are used in such hosts [4].

3.1.4.3 Organic Hosts for Energy Transfer

To avoid quenching effects that inhibit laser emission, it is possible to disperse the emitting molecules in an active host (as opposed to the optically passive polymeric host described above). In this case, the host absorbs the pump energy, and transfers it to the guest by a Förster energy transfer process described in the previous chapter in Sect. 2.4.1. The transfer can be complete even for low doping rates (typically 1 %), which has the additional advantage of reducing the reabsorption of the emitted photons.

An archetypal host-guest system for OLEDs and organic lasers is the combination of Alq_3 as a host and DCM as a guest. The very good overlap in the green part of the spectrum between the DCM absorption and the Alq_3 emission leads to a very high Förster radius of about 4 nm. By using an organic semiconductor as host matrix, such as Alq_3 but also CBP or NPB [61], the emission of the dye can be obtained upon electrical excitation.

3.1.4.4 Organic Liquid Crystals

Liquid crystals are a state of matter that has properties between those of a conventional liquid and those of a solid crystal. They mainly consist of organic molecules, and in this respect can be used as emitters or matrices to host organic laser dyes. In the case of chiral liquid crystals, the helical structure of the nematic liquid crystal provides a periodic variation of the refractive index along the direction perpendicular to the uniformly aligned molecules an consequently plays the role of a DBR cavity (see Sect. 4.2.1.2), which allows easy tuning (see Fig. 3.6) [62, 63]. It has also been shown that for standard thin film laser configurations, the use of a well-oriented liquid crystalline conjugated polymer can

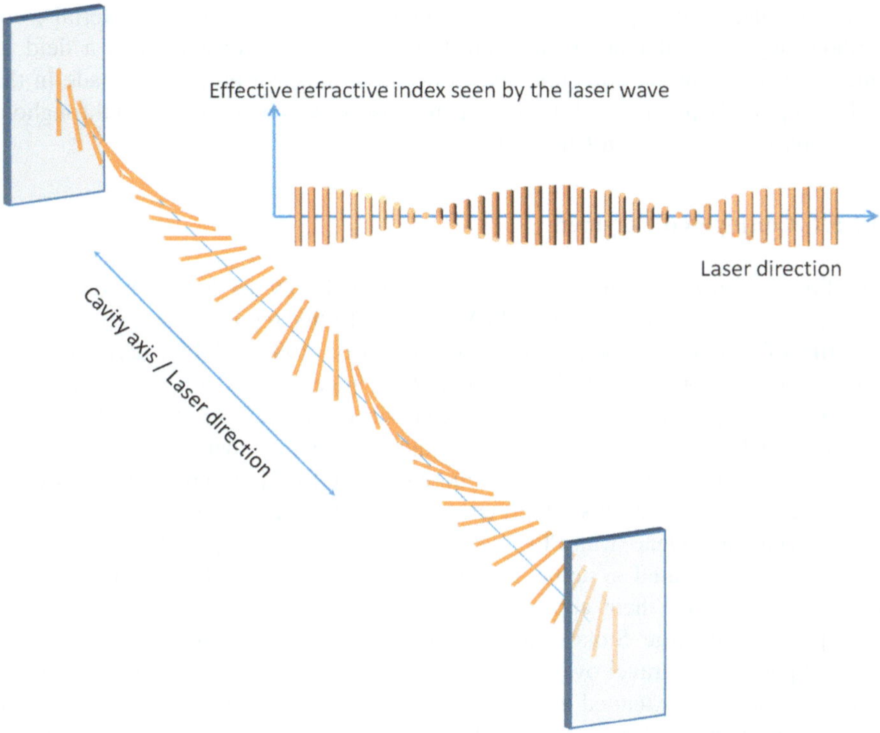

Fig. 3.6 Liquid crystal with periodic index modulation to produce a tunable wavelength through Bragg geometry

improve the optical gain when compared to isotropic materials because all molecules are able to contribute to the laser emission (all the chromophores moments of transition being aligned).

3.1.5 Organic Semiconductors

Organic dyes have been used in laser systems since 1966. However, the renewal of the organic laser field at the end of the twentieth century is mainly attributable to the discovery of semiconducting organic materials, starting with polyacetylene in 1977 [64], by A. Heeger, H. Shirakawa and A. McDiarmid who were awarded the Nobel prize in Chemistry in 2000 "*for the discovery and development of conductive polymers*". This discovery was the starting point of the flourishing field of light emitting organic devices, among which are found organic light emitting diodes [13]. It also launched the still-active quest for the electrically driven organic laser, and indirectly opened new routes for solid-state organic lasers under optical pumping. Optical pumping was primarily considered as an interesting way in the study of fundamental processes and stimulated emission in these materials, but without any particular practical interest; it is now also considered as a field of interest in itself, as we can conclude from the impressive progresses made in the field of optically pumped solid state organic lasers, which are reviewed throughout this book (in particular in Chaps. 4 and 5).

3.1.5.1 Conjugated Polymers

The first demonstration of lasing with a conjugated polymer has been reported by Moses in 1992 in a solution of MEH-PPV [65], but the first reports of laser emission in a neat film of a solid-state conjugated polymer date from 1996 with PPV derivatives in a microcavity [66] or in a mirrorless cavity [67]. A huge number of laser devices based on conjugated polymers have been reported since then [29, 68, 69] with different materials and laser architectures. The conjugated polymers that are mostly used for laser applications are poly(9,9-dioctylfluorene) (PFO) [33], polyphenylene vinylene (PPV) [29, 31], poly(9,9-dioctylfluorene-co-benzothiadiazole (F8BT) [70] or ladder-type polyparaphenylene (MeLPPP) [29] which are shown in Fig. 3.7.

Like all conjugated systems, conjugated polymers are efficient materials for lasing in virtue of their approximate 4-level electronic structure and intense absorption. The large Stokes shift generally observed is also favorable as the emitted photon can travel over relatively long distance without being re-absorbed. Conjugation is not extended over the whole polymer chain but consists in different chromophore units along a given chain. Hence the intermolecular phenomena described in Sect. 2.4 can become *intrachain* phenomena in conjugated polymers (for instance TTA can involve two triplets on a same chain).

The photophysics in conjugated polymers can be somewhat more complicated than in small molecules, as optical excitations can also create polaron pairs, i.e. create an exciton which is splitted into charges which will travel along the polymer chain [72].

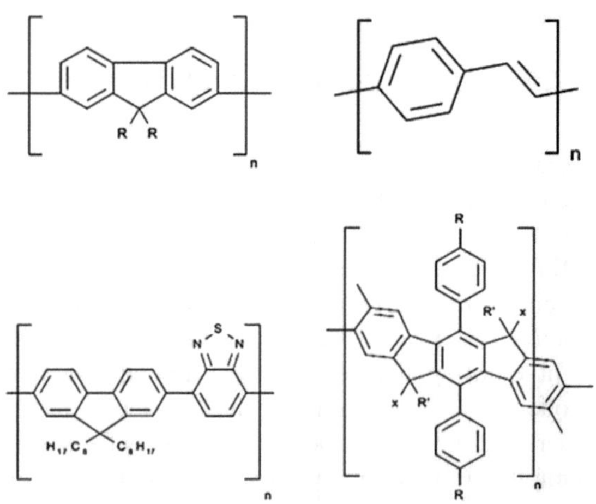

Fig. 3.7 Chemical structure of conjugated polymer materials: poly(9,9-dioctylfluorene) (PFO) (*top left*), polyphenylene vinylene (PPV) (*top right*), poly(9,9-dioctylfluorene-co-benzothiadiazole (F8BT) (*down left*) and ladder-type polyparaphenylene (MeLPPP) (*down right*)—taken from [71], reprint with permission

These differences in photophysics lead to some differences in laser physics: as intermolecular phenomena are more pronounced than in dispersed media (as chromophore units are closer), the dynamics of laser buildup can be affected. For instance, it is observed in a F8BT/MEH-PPV polymer blend that laser pulses do not last more than 10 ns (see Sect. 2.6.3 and Fig. 2.23) while it can be around 100 ns in a dye-doped polymer medium (105 ns in a Rhodamine 6G dye-doped 2-hydroxyethyl methacrylate:methyl methacrylate [73]).

But although these annihilation processes are no doubt more pronounced in neat laser media, conjugated polymers keep a remarkably weak tendency towards concentration quenching, which is the precise reason why they have been so much developed in light-emitting applications. A complete absorption of pump light can be achieved over distances as small as $\sim$200 nm, with still high quantum yields >0.5. Strong absorption and low reabsorption (large Stokes shift) are good features to make lasers in waveguide configuration; that is why this is the preferred laser configuration in which conjugated polymers are used (see Sect. 4.2.1).

Research on conjugated polymers as laser media is motivated by their potential for electrically-driven laser sources. However, serious issues have to be faced before achieving this goal, as explained in details in Chap. 5. Conjugated polymers are nevertheless also attractive under optical pumping as they exhibit very high gains and very low lasing thresholds [69].

3.1.5.2 Organic Crystals

Single crystalline organic semiconductors, or organic crystals are interesting for lasing issues because of their long-range order and high chemical purity. These properties make them potentially able to support high current densities needed to reach lasing threshold under electrical pumping, and to efficiently inject and carry electrical charges (the electron mobility in anthracene is around 1 $cm^2/V.s$, much higher than in PPV).

Moreover, clean cleaved facets can be obtained thanks to the crystalline structure, and even DFB single crystal lasers can be fabricated through laser interference ablation [74].

3.1.6 What is a Good Laser Material?

It is possible to make a short list of the qualities that are required to make a good organic laser material: they pertain both to the fluorophore alone and to its environment (the host matrix for instance).

- Compounds are stable against oxygen and moisture, and photostable against pump photons (pumping at low photon energies greatly relaxes this constraint);
- The quantum yield of fluorescence, measured in the solid state, is high (weak π-π stacking, weak IC and ISC rates, etc.);

- Losses at the laser wavelength are minimized (low reabsorption of the fluorophore and low absorption/scattering losses of the environment);
- The Stokes shift is not too high (reduces the fraction of pump energy converted into heat, and thus enables to increase the conversion efficiency and photostability. However a too small Stokes shift implying an overlap between absorption and emission bands is not good either, see the previous point);
- For a low threshold laser, stimulated emission cross section σ_{em} should be high (governs the local gain $g = \sigma_{em}\,\Delta N$, and then the pump threshold); Note that this is probably not the most important requirement, because all organic emitters with π–π * transitions will have cross sections of the same order of magnitude (see Table 2.1);
- A combination of low intersystem crossing rate, low triplet–triplet absorption cross section, low spectral overlap of optical gain with triplet absorption, and eventually low triplet state lifetime is desirable : all these parameters are often not known but they strongly limit the repetition rate and pulse duration obtainable from optically-pumped organic lasers as discussed in Sect. 2.6;
- In host-guest systems based on Förster resonant energy transfer, a large Förster radius of several nm is desirable (good overlap of donor emission with acceptor absorption, high cross sections…).

3.2 Materials Characterization

In order to qualify a polymer or a dye for laser applications, the knowledge of its basic photophysical properties (absorption and fluorescence spectrum essentially) is a mandatory step but is not enough. Measurement of more advanced parameters like triplet state absorption cross sections, intersystem crossing rate, or triplet lifetime can be useful for the understanding of dynamics, but are not very simple. Most of the time, characterizing a laser material consists in showing that stimulated emission is possible in this particular material (that is, it is possible to reach a laser threshold for which the gain will surpass all sources of losses in the medium, before photodestruction occurs), and quantify these laser capabilities through a measure of net optical gain. For thin-film waveguide lasers, the Variable Stripe Length (VSL technique) is the most widely used technique.

3.2.1 Measuring Stimulated Emission: Optical Gain Measurements

In order to certify a given material for lasing applications, the small-signal gain g can be directly measured, defined by:

$$\frac{1}{I}\frac{dI}{dz} = g = \sigma_{em}\Delta N = \sigma_{em}S_1N$$

where I is the laser intensity in photonic units (photon flux in $m^{-2}\ s^{-1}$) and ΔN is the population inversion (m^{-3}) between the upper level and the lower level of the lasing transition. Typical gains measured in organic gain media are in the range 10^0–10^2 cm^{-1}. In conjugated polymers, transient femtosecond pump/probe experiments revealed gains that can locally rise up to 2000 cm^{-1} in MeLPPP [75] or 12000 cm^{-1} in Polyfluorene [76].

3.2.1.1 The Variable Stripe Length Technique

The Variable Stripe Length (VSL) technique is a widespread technique commonly used to retrieve gain in materials under the form of thin films where a confined mode can propagate in-plane. It consists in measuring Amplified Spontaneous Emission (ASE) in thin-film devices when light can be waveguided inside the active material.

The technique consists in pumping the material from the top by a pump beam whose shape is tailored under the form of a thin stripe of variable length (Fig. 3.8). Spontaneous emission serves as the initial seed for stimulated emission: when net gain is present in the film, light emerging from the edge of the stripe is the result of the fluorescence amplified preferentially along the axis of the stripe; it exhibits a spectrum much narrower than the fluorescence spectrum, often referred to as "mirrorless lasing". Besides, the edge intensity varies superlinearly with the stripe length, and simple data processing allows extracting the net gain [17, 77, 78].

The VSL technique is universal and has been used in thin crystal slabs, organic films, quantum dots, and even silicon [79]. The advantage of this technique relies in its simplicity since it does not require the measurement of other related quantities, and since the injection of a probe beam by the edge of the film is not required (this is a complex issue in organic films where boundaries are never well-defined). The net gain measured this way is a *modal* gain that includes the losses of the waveguide and depends on the geometry of the slab (thicknesses, indexes of refraction). Interestingly, organic lasers are often based on a similar planar slab structure with the addition of gratings to form the resonator (DFB or DBR lasers, see Chap. 4):

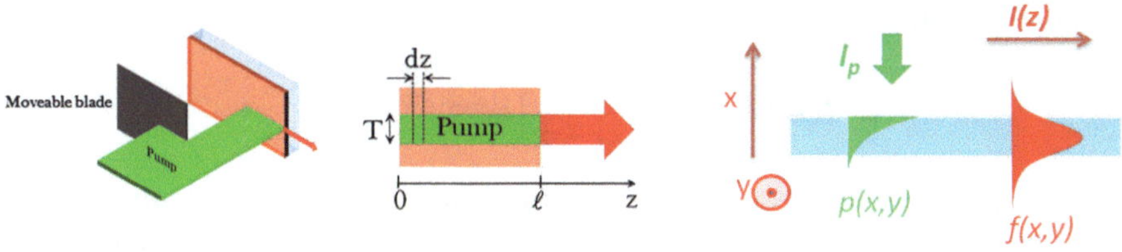

Fig. 3.8 (*Left*) VSL method representation, (*center*) close-up over the pumped region from a view normal to waveguide surface; (*right*) : side view with the pump profile p(x,y) and the mode profile f(x,y)

the gain measured by the VSL technique is then directly the gain present in the final laser system.

Let's consider a thin film of a gain material deposited on a transparent substrate with a lower refractive index to form a planar slab waveguide. If a pump stripe with a homogeneous intensity profile of width T and a variable length ℓ ($T \ll l$) is incident on such a structure, ASE will take place within the excited stripe. The local gain g_{local}, that is, the gain at a given position $\vec{r}$ is given by

$$g_{local}(\lambda, \vec{r}) = \sigma_{em}(\lambda)\Delta N(\vec{r})$$

where σ_{em} is the emission cross section and ΔN represents the inverted population in a given infinitesimal region of the gain medium at $\vec{r}$. As the length of the pump stripe and/or the pump intensity increases, the edge-emitted ASE intensity will grow exponentially. The ASE intensity $I(\lambda, z)$ is a function of z and can be written as:

$$\frac{dI(\lambda, z)}{dz} = [g_{mod}(\lambda, z) - \alpha]I(z) + A(\lambda)I_p$$

where α is the loss coefficient, g_{mod} is the modal gain ($g_{mod} - \alpha$ is the *net* gain) and $A(\lambda)I_p$ is the seeding term, the one responsible for the process to start, related to the power density of the spontaneously-emitted photons into the stripe solid angle. It is written under this form to point out that spontaneous emission is a linear phenomenon proportional to the pump intensity and has the same emission spectrum as stimulated emission, i.e. $A(\lambda) \propto \sigma_{em}(\lambda)$.

The modal gain $g_{mod}(\lambda)$ in the z-direction is a spatial average over the cross section S of the waveguide, of the local gain, weighted by the pump energy distribution $p(x, y)$ (if the pump is incident from the top, as sketched in Fig. 3.8, the pump energy density follows an exponential decay along the x direction) and by the mode profile distribution $f(x, y)$:

$$g_{mod}(\lambda, z) = \iint_S g_{local}(\lambda, \vec{r})f(x, y)p(x, y)dS$$

If the modal gain is supposed to be constant with z (in practice as the pump stripe has a Gaussian dependence along z, the stripe must then be made well longer than the domain of variation of z during the VSL measurement), and if gain saturation is neglected, integrating the equation from z = 0 to z = L with the assumption that I(0) = 0 yields:

$$I(\lambda, L) = \frac{A(\lambda)I_p}{g_{mod}(\lambda) - \alpha}\left[e^{(g_{mod}(\lambda)-\alpha)L} - 1\right]$$

This simple model also assumes the pump stripe is almost one dimensional. This means that because $T \ll l$ stimulated photons will emerge preferentially along the stripe direction.

The most striking manifestation of ASE is a narrowing of the emission spectrum, as the fluorescence photons at the edges of the emission spectrum are much

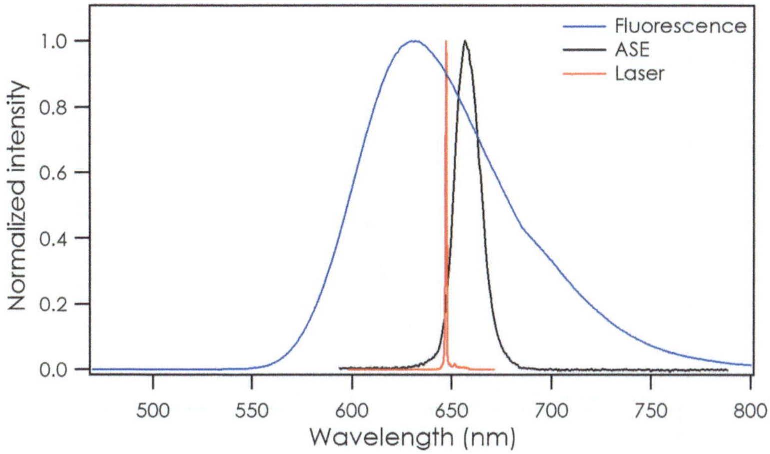

Fig. 3.9 Fluorescence, amplified spontaneous emission and laser spectrum in a thin-film laser consisting of a neat film of small molecules [17]. ASE is characterized by a thinner spectrum than fluorescence (typically ~10 nm), but in general much broader than laser emission. (Reprinted with permission from [17]. Copyright 2009, American Institute of Physics)

less amplified than those around the maximum. The ASE spectra are still quite broad though (typically a few nm) and usually broader than laser emission, see Fig. 3.9. The other peculiarity of ASE, as seen in the previous equation, is its superlinear dependence with L, in contrast with fluorescence which varies linearly with L. This last feature enables to make a clear distinction between ASE and filtered fluorescence or spectrally-narrow waveguide emission originating from the existence of leaky modes for instance [80].

Neglecting gain saturation (see Sect. 1.4) is a strong assumption which in many practical cases is not justified. An example of the gain saturation effect is shown in Fig. 3.10.

VSL is a very powerful technique: no specific sample preparation is required, and the model (one-dimensional) is extremely simple. However, several limitations can lead to erroneous values of the measured gain, as described in detail in [77]. First of all, special care has to be paid at the gain saturation issues: for very high gains, saturation occurs for very short stripes, where the 1-D model is not valid anymore. In addition, because organic materials have both very high gains and optical scattering properties (films are usually not optically perfect, dust or point defects are present on the surface, etc.), they are easily prone to random lasing effects, which can compromise a correct measurement of ASE (see Sect. 4.2.1.3).

Other experimental flaws can arise from the non-uniform pump intensity profile, the coupling efficiency to the detector (which should not depend on the stripe length) or the diffraction at the edge of the razor blades. A simple way to limit the diffraction effects is to realize the variable stripe by imaging a couple of razor blades (see Fig. 3.11) instead of placing the sample at a fixed distance, which virtually eliminates the Fresnel diffraction pattern and ensures a sharp profile for

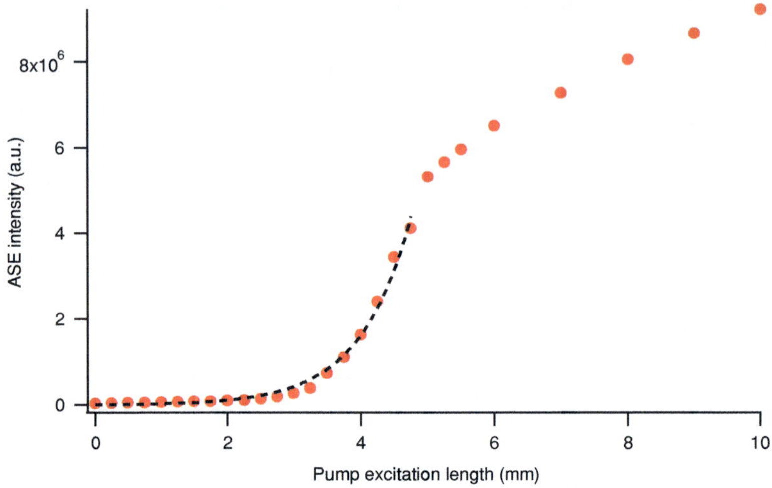

Fig. 3.10 Gain saturation demonstration in ASE collected in a VSL experiment. Here, a thin film of PMMA doped with rhodamine 640 (from the authors). The *dashed line* is a fit to the experimental data by using the classical VSL expression. Gain saturation is the most likely explanation for the departure of the data from simple VSL theory; problems due to signal collecting also arise at long excitation lengths as it becomes difficult to collect all the edge-emitted light without lateral confinement in the waveguide

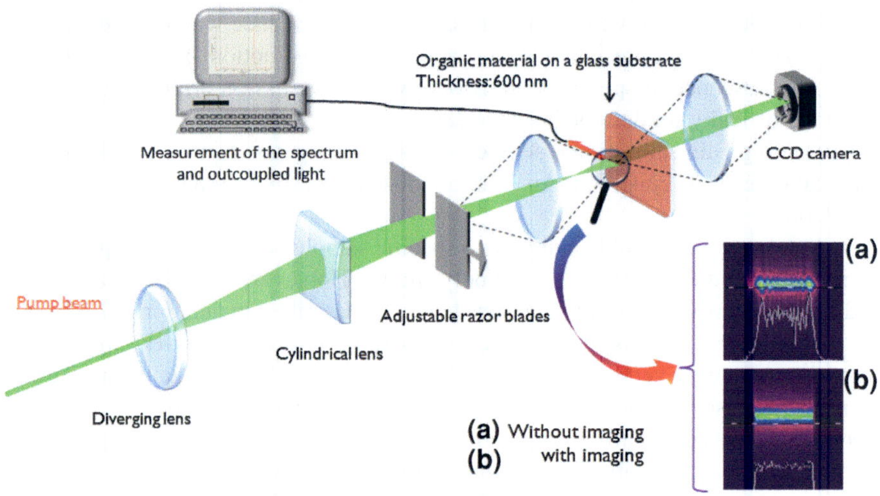

Fig. 3.11 VSL setup with imaging system between the razor blades and the active medium to eliminate the complex Fresnel diffraction pattern onto the film (see inset, case **a** without imaging, **b** with imaging)

the pump stripe (provided that the numerical aperture of the imaging lens is high enough to collect the high-spatial frequency contents of the blades diffraction pattern), thus eliminating the need of a complex algorithm to retrieve the gain [17].

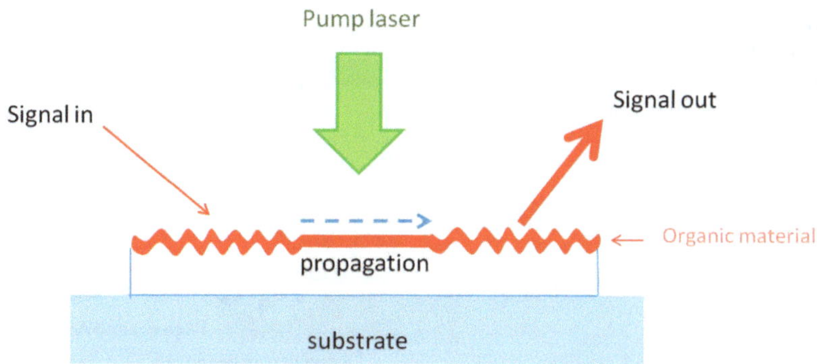

Fig. 3.12 Use of DBR grating to couple the light in and out of a pumped organic dye-doped or polymeric waveguide. The optical gain can be obtained from the ratio P_{out}/P_{in}

3.2.1.2 Other Gain Measurement Techniques

As stated in the previous paragraph, the VSL technique suffers from some drawbacks, especially when high gains have to be measured. In this case, other techniques can be implemented such as transient absorption: the vertical (thin) dimension of the organic film is used, so that propagation over at most ~ 1 μm cannot enable intensities to reach the saturation intensity. The organic sample is illuminated with a short laser pulse, and a probe pulse is sent with a controlled time delay to probe the optical gain.

Another simple setup consists in the injection of a small probe beam in a planar waveguide and measuring the outcoupled beam after travelling through a pumped zone of a given length. A technical trick is needed here because the injection of a laser beam in a typically 100-nm thick polymeric waveguide with poorly defined end-facets is not an easy task. A simple solution is to engrave on the polymer sample two Bragg gratings, one is used to efficiently couple the light inside the waveguide, the other being for coupling the amplified light out. The grating axis are not aligned one to another to avoid parasitic lasing. For example, with this method, optical gain in a MEH-PPV sample as high as 100 mm^{-1} was measured [81] (Fig. 3.12).

3.2.2 *Loss Measurements*

The gain coefficient measured through the VSL technique is the net gain $g_{mod} - \alpha$ which is the gain minus the waveguide propagation losses. These losses might be due to scattering, reabsorption losses at the emission wavelength, or excited state absorption, especially Triplet absorption for long pump pulses.

The ensemble of the above-mentioned losses can be determined through a modified VSL technique. In this technique the sample is pumped with a stripe of

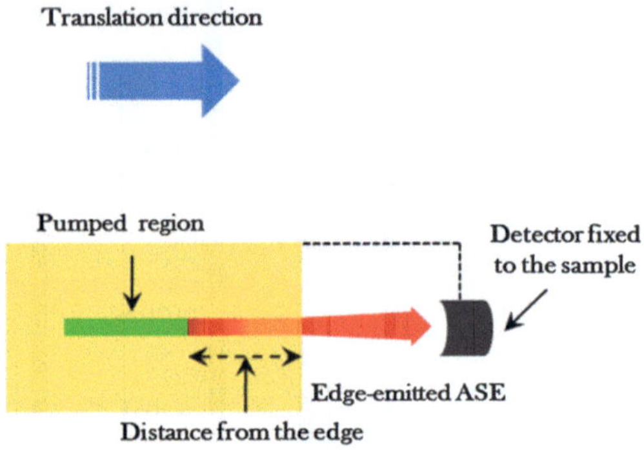

Fig. 3.13 Waveguide propagation losses measurement technique derived from the VSL method

constant length and the edge-emitted intensity is monitored while the sample and detector, fixed one with respect to each other, are translated laterally in the normal direction with respect to the pump beam. This resembles the condition when the pump stripe gradually moves away from the edge of the sample (Fig. 3.13). The edge-emitted ASE emission decreases since it has to travel across an increasing unpumped length before reaching the sample edge. Keeping the shape of a stripe, instead of a point source, is essential to create ASE and not simple fluorescence: it makes a quite directional beam in virtue of gain guiding inside the pumped region, and ASE is furthermore much more intense than simple fluorescence. Assuming a collimated beam means that all the light emerging from the edge must be properly collected, so that the decrease of received light with distance might be only ascribed to losses.

In this ideal case the ASE intensity depends on the length x of the unpumped region between the stripe and the edge as:

$$I_{out} = I_o exp(-\alpha x)$$

With I_0 the intensity at the end of the pump stripe and α is the waveguide loss coefficient.

Loss coefficients measured by this technique are very approximate values and precautions must be taken when using these quantities. In fact, several factors are responsible for considerable uncertainty of this method. The first problem is the diameter of the detector that must be larger than the pump stripe diameter to collect all the edge-emitted light. This puts a limit on the diameter of the pump stripe. The second problem is that in amorphous thin films, the edges are usually very scattering for light, even after a clean sample cleavage. As the pump stripe moves away from the sample edge, the solid angle becomes wider and thus the fraction of the edge-scattered light, seen by the detector, is varying for each distance.

Given the above limitations, depending on how the detector is positioned at the sample edge, the edge emitted intensity would be different from one measure to the other, leading to different loss coefficients for a given waveguide structure.

Therefore, an alternative method of loss measurement can be useful. It is for example possible to determine the loss coefficient by imaging (from above the sample) the guided light scattered from the film surface. As the light propagates along the pump stripe inside the medium, the intensity of the guided light will decay exponentially from the end of the pump stripe (Fig. 3.15). Therefore, if the guided light intensity profile is imaged and fitted by an exponential, it can determine the waveguide loss coefficient in a precise manner. In this way, the problems related to the light collection method can be resolved and the loss measurement will be further reliable and repeatable [82, 83] (Fig. 3.14).

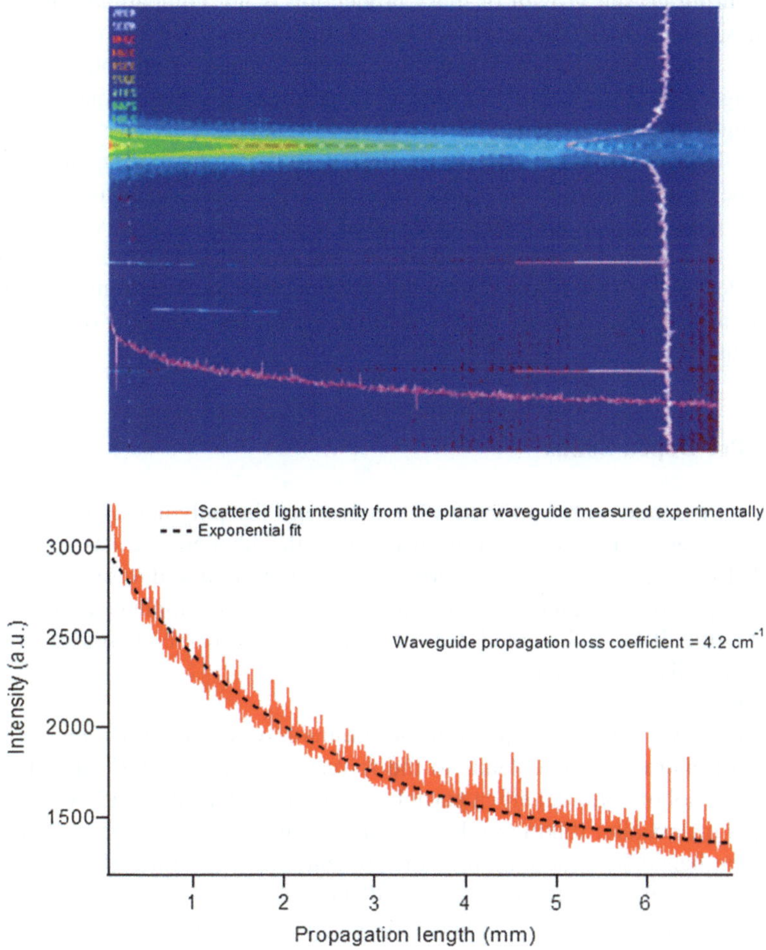

Fig. 3.14 Waveguide loss measurement on a 6-μm thick film of PMMA doped with the classical Rhodamine 640 laser dye (doping rate: 1 % of the total solid matrix weight) spin coated onto the silica substrate. (*top*) image of the guided light along the pump stripe obtained by CCD camera. (*bottom*) the intensity profile of the guide light. The dashed line is an exponential fit to the experimental data.

3.3 Fabrication of Organic Lasing Gain Chips: Bulk Rods and Thin Films

3.3.1 Bulk Rods

Maybe the most direct way to make an organic solid-state laser is to take a dye molecule and insert it directly as a dopant into a solid. Inorganic crystals or glasses would be ideal hosts for organic molecules, they have good optical and mechanical properties and can be easily polished, but the high temperatures (around 1000 °C) needed during crystal growth or glass production is not compatible with organic molecules, which would be destroyed during the process. There are consequently two main others approaches to obtain a piece of glassy (amorphous) material containing organic dyes: the first one is to produce this glassy material at low temperature, in order to obtain either a purely organic (a polymer) or an organic/inorganic hybrid; the second one is to use a porous inorganic glass and dope the dye inside it after fabrication, upon impregnation. In the first case, the optical quality is good but the material is somewhat softer than silica glass for instance, making it more difficult to shape and polish, and less resistant to very high optical powers. Post-doping leads to a solid material very close to glass from a mechanical point of view, but the porosity could lead to diffusion and generally offers a poorer optical quality.

3.3.1.1 Pre-Doped Solids

A solution is made with a mixture of the dye and the polymer used for the host matrix. During all the polymerisation process, the temperature has to be maintained below a few hundreds of degrees to keep the organic dye safe. Dye-doped sol-gels or poly-(methyl methacrylate) (PMMA) can for example be produced with this technique. It has to be noted that the curing process can be long (up to several weeks, typically several days for PMMA).

Sol-gels

The hydrolysis at low temperature of tetramethoxysilane with water in methyl alcohol (or another solvent) leads to the formation of the matrix. After polycondensation, a sol of colloidal silica particles in suspension is obtained, which lead to a rigid gel matrix after linking of the sol-particle together. This "xerogel" matrix is then strengthened by evaporation of the solvent and water through a drying process. The xerogel structure consists in a 3D network of 1 to 10 nm large pores, in which dye molecules can be linked through hydrogen bonding with the silanol groups on the pore surface [7].

PMMA

Transparent (nonconductive) polymers have been used as hosts for dye molecules in early solid-state laser devices, because of their excellent optical quality (transparency, homogeneity) and their simple and well-controlled fabrication process.

The host matrix comes here from the polymerization of the monomer methyl methacrylate, in which the organic dye is dissolved. The reaction may take a few days, under controlled low temperature (around 50 °C). PMMA is the archetypal polymer used as a host matrix, but modified PMMA (for instance copolymer such as methyl methacrylate/hydroxyl ethyl methacrylate) is also used to enhance the damage threshold or to use specific dyes (for example, rhodamine 6G is not soluble in methyl methacrylate) [3, 84].

3.3.1.2 Post Doped Solids

The post-doping method consists in introducing a dye solution into a glass-like structure [85]. The main advantage of this kind of solid is their mechanical properties, very close to glass because a high temperature process can be used. The glass has then to be made porous, either chemically or by etching, and impregnated with the chosen dye. For example, a microporous glass can be obtained through selective etching of sodium borosilicate glass [85]. The exact amount of dye inside the matrix, as well as the doping homogeneity, is however difficult to evaluate. This measurement is often done after the impregnation, upon evaluating the difference between the amount of solution after and before the doping process. The lack of accuracy of this method can be very problematic for some dyes which are very sensitive to concentration quenching for example: in this case the level of doping has to be very low (1 % or less) and controlled very precisely.

After impregnation, the sample has to be polished with optical quality to be inserted in the laser cavity without adding important losses. This step is a technical challenge due to the porous morphology of the matrix. It is anyway difficult to avoid a certain level of scattering losses inside the bulk material due to the pores themselves.

3.3.2 Thin-Films

Organic materials can be easily shaped as thin films with thicknesses from a few nanometers to a few micrometers with relatively easy, straightforward and low cost methods. Numerous techniques exist to form such thin films, including thermal evaporation, spin coating, inkjet printing, plasma enhanced chemical vapour deposition, pulsed laser deposition, electrospray…. In the following, the main methods used to shape solid-state organic lasers will be briefly described.

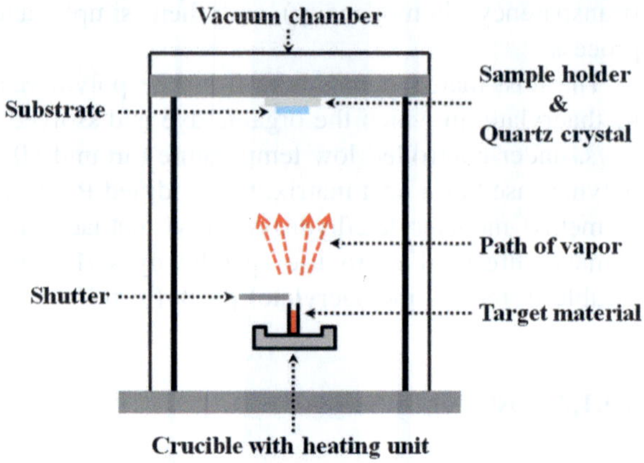

Fig. 3.15 Thermal evaporation process

3.3.2.1 Thermal Evaporation

Thermal evaporation or sublimation is a well-known and widely used technique to deposit thin films of low-weight organic molecules. In particular, this method is used by the OLED industry to realize complex organic heterostructures. The process takes place in a vacuum chamber (around 10^{-6}–10^{-7} mbar): organic molecules, placed in crucibles at the bottom of the chamber, are heated by an electric resistance until sublimation and are condensed back on the substrates placed on top of the chamber (Fig. 3.15). The speed of evaporation is linked to the crucible temperature, and controlled in real-time by a quartz microbalance. Very accurate thickness control can be achieved (<1 nm) and multilayer deposition is easy to perform. Simultaneous sublimation of different materials at different rates can also be done to make host-guest systems.

Thermal evaporation is a powerful technique allowing the fabrication of complex multilayer structures with high-optical quality and well-controlled thicknesses. However, simpler solution-based techniques are often preferred because sublimation is not possible with heavy molecules such as polymers, and requires a costly vacuum chamber.

3.3.2.2 Solution-Processing: Spin Coating, Dip Coating, Doctor Blading, Printing

Deposition from solution is the main fabrication technique used for organic lasers. The principle is simple: a solution of the active medium (a conjugated polymer, or a transparent polymer doped with a laser dye,...) is first prepared in an organic solvent and deposited onto a sample. The control of film thickness, optical quality and homogeneity will depend on the choice of the solvent and the choice of the concentration of the laser material in the solvent, as they play on the viscosity and on the wetting properties of the liquid, which ultimately control the film properties.

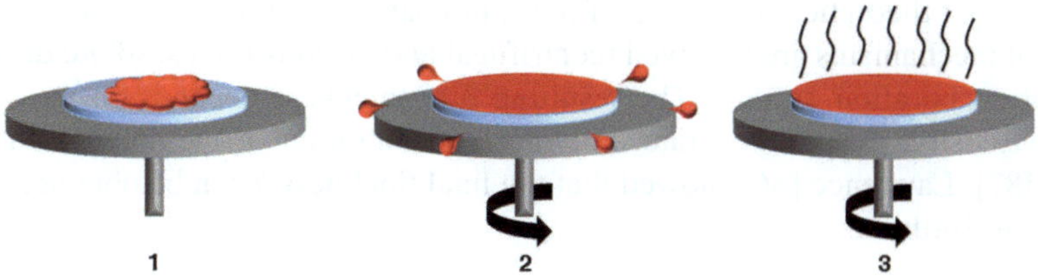

Fig. 3.16 Spin coating process: a drop of polymer solution is deposited onto the substrate (*1*), then the fast rotation of the substrate ejects the excess of polymer solution and give birth to a nearly homogeneous thin film (*2*). Heating the sample allows evaporation of the remaining solvent (*3*)

Spin Coating

In this case, centrifugal force is used to create high-quality thin films. The substrate is placed on the spin coater vacuum chuck and is vacuum clamped for spinning. In order to assure wetting of the whole substrate area, an excess amount of the polymeric solution compared to what is finally required is dispensed on the center of the substrate. As the spin coater speeds up according to a predefined program, the centrifugal acceleration causes the solution to spread up to the edge of the substrate, leaving a thin film of solution on the surface. The samples obtained by this technique are transferred to the oven and annealed to evaporate the solvent and to make the film denser. Figure 3.16 is schematically showing different steps in spin coating process.

For a well-known polymer such as PMMA with a given viscosity, the control of the thickness through the rotation speed of the substrate is very accurate. However, as other solution-processing techniques, it is difficult to coat several superimposed layers, because solvents need to be non-miscible.

Edge-effects are also responsible for a non-homogeneous thickness near the edge of the substrate (Fig. 3.17), especially for high-viscosity solutions.

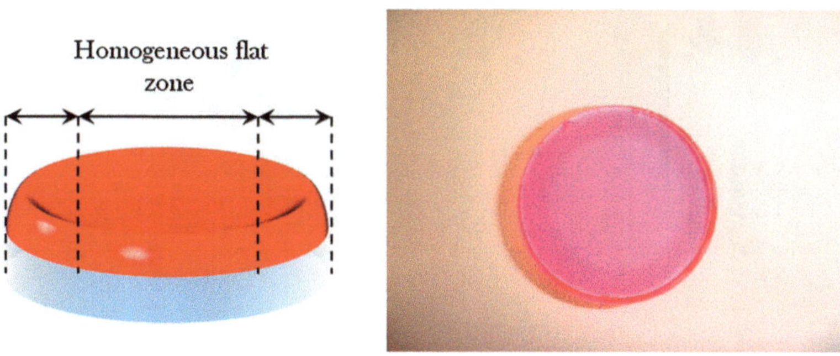

Fig. 3.17 For high-viscosity polymers, edge-effects are visible and an increase of the thickness is observed at the edge of the substrate, resulting in a darker (more absorbing) ring on the border on the photography of this Rhodamine 640-doped PMMA on a glass sample

The exact theoretical modeling of the spin-coating process is complex, as many physical mechanisms are involved (centrifugal and viscous forces, solute diffusion, solvent evaporation…) [86]. The resulting film thickness depends on the solvent type, viscosity, evaporation rate, polymer concentration in the solvent and spin speed [87]. Lawrence [86] showed that the final thickness h can be obtained by the following formula:

$$h = k \cdot C_0 \cdot (\eta \cdot D)^b \omega^a$$

where k is a constant (around unity), C_0 is the initial polymer concentration, η is the initial solution viscosity, D is the solute diffusivity and ω is the angular velocity. a and b are empirically determined constants for the power laws, which only slightly vary with the type of polymer used (a is around -0.5 and b around 0.35). The value of the viscosity and diffusivity are however very hard to find for a given polymer, making the above formula of little practical interest.

However, a great deal of work has been reported to deduce the polymer final thickness from experimental simple parameters such as angular velocity and solution viscosity. It has been proven that some other parameters (amount of solution deposited on the substrate at the beginning of the process, acceleration ramps before the final rotation speed is reached, total spin time) have very limited effects on the final film thickness. For a given polymer in a given solution, the manufacturer is generally able to provide a very accurate curve describing the final thickness obtained for a given angular velocity of the rotating disc.

Dip Coating

Dip coating techniques can be described as a process where the substrate to be coated is immersed in a liquid and then withdrawn with a well-defined withdrawal speed under controlled temperature and atmospheric conditions (Fig. 3.18). The

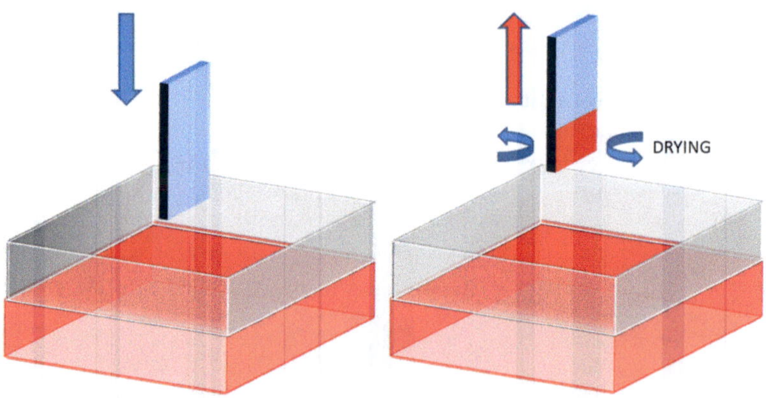

Fig. 3.18 Dip-coating process. The substrate is dipped into the solution, then slightly removed for drying

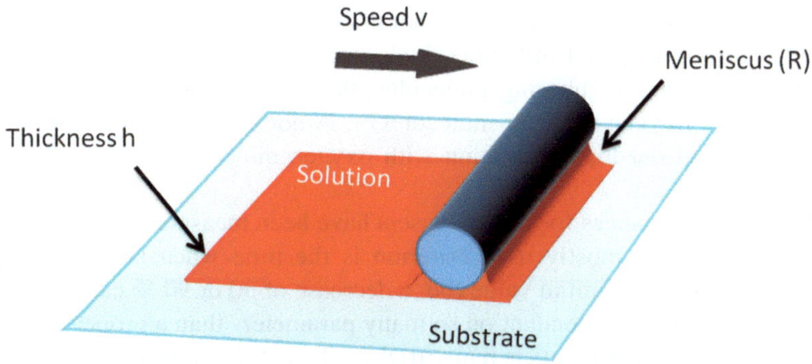

Fig. 3.19 Schematic drawing of the horizontal dipping process

coating thickness is mainly defined by the withdrawal speed, by the solid content and the viscosity of the liquid.

Horizontal Dipping

There are several slightly-different techniques consisting in dipping a solution horizontally between a substrate and a blade or rod, called "horizontal dipping", "Doctor blading" [42], "wire-bar coating" [88].

The apparatus for horizontal dipping consists in a cylindrical rod parallel to the surface to coat, and located at a controllable distance h from the surface. This gap (h) is filled with the organic solution with a syringe to form a meniscus (Fig. 3.19). The cylinder is then moved, through an automated microtranslation system parallel to the substrate surface at a defined speed. The film thickness is given by [89]:

$$d = 1.34 \cdot R \cdot (\mu v/\sigma)^{2/3}$$

where μ is the viscosity, σ the surface tension, v the translation speed and R the radius of curvature of the meniscus. As a consequence, the film thickness can be controlled by varying the translation speeds. With this method, large area homogeneous films can be produced. By increasing the translation speed during the process, it is even possible to obtain wedged organic layers with a continuously varying thickness all over the sample. This kind of layer can be useful to tune the lasing wavelength by modifying the effective index in a DBR laser [89].

3.4 Photodegradation Issues in Organic Lasers

Photo-degradation issues are maybe the most problematic ones when dealing with organic molecules. The associated mechanisms are very complex and no complete understanding has been reported up to now, despite numerous studies [90–93] on the subject. The photochemical and photothermal properties of dyes are indeed dependant of a huge panel of factors, including the structure of the dye itself, its interaction with the neighboring molecules, the external environment (especially the presence of oxygen or water molecules)… Among the parameters involved, it has been pointed out that interaction with oxygen molecules is very detrimental [90].

Lifetime performances for organic lasers have been measured and reported with various criteria. The mostly used criterion is the time when laser intensity has decreased by half of its initial value (but a decrease of 30 or 90 % can sometime be used). This lifetime is dependant on so many parameters than a proper comparison between different experiments is really tricky. The nature of the dye, of course, but also the matrix, the excitation wavelength, intensity, pulse duration and repetition rate, the resonator configuration, as well as the atmosphere used during fabrication and/or lasing measurement (ambient conditions, encapsulating, in a glove box, in vacuum…) are all playing a crucial role on the observed lifetime of the device. Obviously, the best performances are obtained with encapsulated systems [91, 92] (or under vacuum), short pulses, low repetition rates and optimized choice of the dye/matrix couple [5, 93, 94].

As a consequence, only rough orders of magnitude can be given, as well as some enhancement factors obtained in carefully designed experiments where only one parameter is varied.

For bulk solid-state laser, encapsulation is not an option, and experiments are usually conducted in ambient conditions. One of the best dye in terms of photostability is pyrromethene, and several studies were made to compare different modified PMMA matrices from the lifetime point of view [3, 95].

Typical lifetimes for bulk solid-state lasers are on the range of 10^6 pulses.

For thin film lasers, encapsulation is possible and benefits from the work made in this direction by the OLED community. It has been demonstrated by Richardson et al. [91] that a 2500 fold increase in lifetime can be measured for a polymer DFB laser if the polymer is protected from oxygen by an optical adhesive. The lifetime of the laser is above 10 millions of pulses.

Of course, these figures are given for a single pump spot on the organic material. A very natural and simple way of increasing the lifetime of organic laser devices is to rotate and/or translate the lasing medium to address fresh spots at regular time intervals. Depending on the size of the lasing material and on the required stability (a rotating disc will for example induce short time noise [96]) a translation of the sample with respect to the pump spot can be made after a decrease of the laser intensity by a given value, leading to an increase of the operational lifetime of the laser.

References

1. F.P. Schafer, W. Schmidt, J. Volze, Organic dye solution laser. Appl. Phys. Lett. **9**(8), 306 (1966)
2. B. Valeur, *Molecular Fluorescence* (Wiley-VCH Verlag GmbH, Weinheim, 2001)
3. A. Costela, I. García-Moreno, R. Sastre, Solid-state dye lasers, in *Tunable Laser Applications* 2nd edn., ed. by F.J. Duarte (CRC, New York, 2009), pp. 97–120
4. M. Ahmad et al., Performance and photostability of xanthene and pyrromethene laser dyes in sol-gel phases. J. Phys. D-Appl. Phys. **35**(13), 1473–1476 (2002)
5. M.D. Rahn, T.A. King, Comparison of laser performance of dye molecules in sol-gel, polycom, ormosil, and poly(methyl methacrylate) host media. Appl. Opt. **34**(36), 8260–8271 (1995)
6. E. Yariv, R. Reisfeld, Laser properties of pyrromethene dyes in sol-gel glasses. Opt. Mater. **13**(1), 49–54 (1999)
7. M. Faloss et al., Toward millions of laser pulses with pyrromethene- and perylene-doped xerogels. Appl. Opt. **36**(27), 6760–6763 (1997)
8. A. Moliton, *Optoelectronics of Molecules and Polymers* (Springer, New York, 2005)
9. A. Bernanose, M. Comte, P. Vouaux, Sur un nouveau mode d'émission lumineuse chez certains composés organiques. J. Chim. Physique (in french) **50**, 64–68 (1953)
10. M. Pope, H.P. Kallmann, P. Magnante, Electroluminescence in organic crystals. J. Chem. Phys. **38**, 2042–2043 (1963)
11. B.K. Yap et al., Simultaneous optimization of charge-carrier mobility and optical gain in semiconducting polymer films. Nat. Mater. **7**(5), 376 (2008)
12. S.Z. Bisri et al., High mobility and luminescent efficiency in organic single-crystal light-emitting transistors. Adv. Funct. Mater. **19**(11), 1728 (2009)
13. C.W. Tang, S.A. Vanslyke, Organic electroluminescent diodes. Appl. Phys. Lett. **51**(12), 913–915 (1987)
14. C.W. Tang, S.A. Vanslyke, C.H. Chen, Electroluminescence of doped organic thin-films. J. Appl. Phys. **65**(9), 3610–3616 (1989)
15. E. Ishow et al., Multicolor emission of small molecule-based amorphous thin films and nanoparticles with a single excitation wavelength. Chem. Mater. **20**(21), 6597–6599 (2008)
16. S. Forget et al., Red-emitting fluorescent organic light emitting diodes with low sensitivity to self-quenching. J. Appl. Phys. **108**(6) (2010)
17. H. Rabbani-Haghighi et al., Laser operation in nondoped thin films made of a small-molecule organic red-emitter. Appl. Phys. Lett. **95**(3), 033305 (2009)
18. J.C. Ribierre et al., Amplified spontaneous emission and lasing properties of bisfluorene-cored dendrimers. Appl. Phys. Lett. **91**(8) (2007)
19. G. Jordan et al., Light amplification at 501 nm and large nanosecond optical gain in organic dye-doped polymeric waveguides. Appl. Phys. Lett. **88**(16) (2006)
20. R. Xia et al., Low-threshold distributed-feedback lasers based on pyrene-cored starburst molecules with 1,3,6,8-attached Oligo(9,9-Dialkylfluorene) arms. Adv. Funct. Mater. **19**(17), 2844 (2009)
21. D. Schneider et al., Laser threshold reduction in an all-spiro guest-host system. Appl. Phys. Lett. **85**(10), 1659–1661 (2004)
22. H. Nakanotani et al., Extremely low-threshold amplified spontaneous emission of 9,9'-spirobifluorene derivatives and electroluminescence from field-effect transistor structure. Adv. Funct. Mater. **17**, 2328–2335 (2007)
23. D. Schneider et al., An ultraviolet organic thin-film solid-state laser for biomarker applications. Adv. Mat. **17**(1), 31 (2005)
24. N. Johansson et al., Solid-state amplified spontaneous emission in some spiro-type molecules: a new concept for the design of solid-state lasing molecules. Adv. Mater. **10**(14), 1136 (1998)
25. M. Berggren, A. Dodabalapur, R.E. Slusher, Stimulated emission and lasing in dye-doped organic thin films with Forster transfer. Appl. Phys. Lett. **71**(16), 2230–2232 (1997)

26. V.G. Kozlov et al., Laser action in organic semiconductor waveguide and double-heterostructure devices. Nature **389**(6649), 362–364 (1997)
27. V.G. Kozlov et al., Study of lasing action based on Forster energy transfer in optically pumped organic semiconductor thin films. J. Appl. Phys. **84**(8), 4096–4108 (1998)
28. A.J. Heeger, Nobel lecture: semiconducting and metallic polymers: the fourth generation of polymeric materials. Rev. Mod. Phys. **73**(3), 681–700 (2001)
29. U. Scherf et al., Conjugated polymers: lasing and stimulated emission. Curr. Opin. Solid State Mater. Sci. **5**(2–3), 143–154 (2001)
30. Y.J. Chen et al., Laser action in a surface-structured free-standing membrane based on a pi-conjugated polymer-composite. Org. Electron. **12**(1), 62–69 (2011)
31. W. Holzer et al., Corrugated neat thin-film conjugated polymer distributed-feedback lasers. Appl. Phys. B-Lasers Opt. **74**(4–5), 333–342 (2002)
32. M. Leclerc, Polyfluorenes: twenty years of progress. J. Polym. Sci., Part A: Polym. Chem. **39**(17), 2867–2873 (2001)
33. G. Heliotis et al., Blue, surface-emitting, distributed feedback polyfluorene lasers. Appl. Phys. Lett. **83**(11), 2118–2120 (2003)
34. M. Lehnhardt et al., Impact of triplet absorption and triplet-singlet annihilation on the dynamics of optically pumped organic solid-state lasers. Phys Rev B **81**(16), 165206 (2010)
35. H. Coles, S. Morris, Liquid-crystal lasers. Nat. Photon **4**(10), 676 (2011)
36. F. Muñoz, A.P. Palffy-Muhoray, B. Taheri, Ultraviolet lasing in cholesteric liquid crystals. Opt. Lett. **26**(11), 804 (2001)
37. M.C. Gather, S.H. Yun, Single-cell biological lasers. Nat. Photon **5**(7), 406 (2011)
38. S. Meech, Biophotonics: living lasers. Nat. Photon **5**(7), 387 (2011)
39. I. Garcia-Moreno et al., Materials for a reliable solid-state dye laser at the red spectral edge. Adv. Funct. Mater. **19**(16), 2547 (2009)
40. A. Deshpande, J. Rane, L. Jathar, Comparison of spectroscopic and lasing properties of different types of sol-gel glass matrices containing Rh-6G. J Fluoresc **19**(6), 1083 (2009)
41. C.-C. Chang et al., Spin coating of conjugated polymers for electronic and optoelectronic applications. Thin Solid Films **479**(1–2), 254 (2005)
42. I.K. Ding et al., Deposition of hole-transport materials in solid-state dye-sensitized solar cells by doctor-blading. Org. Electron. **11**(7), 1217–1222 (2010)
43. Y. Yoshioka, G.E. Jabbour, Desktop inkjet printer as a tool to print conducting polymers. Synth. Met. **156**(11–13), 779 (2006)
44. H. Rabbani-Haghighi et al., Laser operation in nondoped thin films made of a small-molecule organic red-emitter. Appl. Phys. Lett. **95**(3) (2009)
45. E.J.W. List et al., Direct evidence for singlet-triplet exciton annihilation in pi-conjugated polymers. Phys. Rev. B **66**(23) (2002)
46. C. Gartner et al., The influence of annihilation processes on the threshold current density of organic laser diodes. J. Appl. Phys. **101**(2), 023107 (2007)
47. E.J.W. List et al., Interaction of singlet excitons with polarons in wide band-gap organic semiconductors: a quantitative study. Phys. Rev. B **64**(15), 155204 (2001)
48. R. Osterbacka et al., Excitons, polarons, and laser action in poly(p-phenylene vinylene) films. J. Chem. Phys. **118**(19), 8905 (2003)
49. B. Kraabel et al., Unified picture of the photoexcitations in phenylene-based conjugated polymers: Universal spectral and dynamical features in subpicosecond transient absorption. Phys. Rev. B **61**(12), 8501 (2000)
50. M. Koschorreck et al., Dynamics of a high-Q vertical-cavity organic laser. Appl. Phys. Lett. **87**(18) (2005)
51. N.C. Giebink, S.R. Forrest, Temporal response of optically pumped organic semiconductor lasers and its implication for reaching threshold under electrical excitation. Phys. Rev. B **79**(7) (2009)
52. T.G. Pavlopoulos, Scaling of dye lasers with improved laser dyes. Prog. Quantum Electron. **26**(4–5), 193–224 (2002)

53. F.P. Schafer (ed.), Dye lasers in *Topics in Applied Physics*, vol. 1, (Springer, New York, 1973). 285
54. M.P. O'Neil, Synchronously pumped visible laser dye with twice the efficiency of Rhodamine 6G. Opt. Lett. **18**(1), 37–38 (1993)
55. F. López Arbeloa et al., Structural, photophysical and lasing properties of pyrromethene dyes. Int. Rev. Phys. Chem. **24**(2), 339–374 (2005)
56. F. Liang et al., Eight-position substitution effects on laser action of the 1,3,5,7-tetramethyl-2,6-diethyl pyrromethene-BF2 complexes. J. Opt. Soc. Am. B **18**(12), 1841–1845 (2001)
57. A. Costela et al., New analogues of the BODIPY dye PM597: photophysical and lasing properties in liquid solutions and in solid polymeric matrices. J. Phys. Chem. A **113**(28), 8118–8124 (2009)
58. S. Mula et al., Design and development of a new pyrromethene dye with improved photostability and lasing efficiency: theoretical rationalization of photophysical and photochemical properties. J. Org. Chem. **73**(6), 2146–2154 (2008)
59. S. Chenais et al., On thermal effects in solid-state lasers: the case of ytterbium-doped materials. Prog. Quantum Electron. **30**(4), 89–153 (2006)
60. A. Costela et al., High-repetition-rate polymeric solid-state dye lasers pumped by a copper-vapor laser. Appl. Phys. Lett. **79**(4), 452–454 (2001)
61. D. Schneider et al., Ultrawide tuning range in doped organic solid-state lasers. Appl. Phys. Lett. **85**(11), 1886–1888 (2004)
62. M. O'Neill, S.M. Kelly, Ordered materials for organic electronics and photonics. Adv. Mater. **23**(5), 566–584 (2011)
63. H. Coles, S. Morris, Liquid-crystal lasers. Nat. Photon **4**(10), 676–685 (2010)
64. C.K. Chiang et al., Electrical conductivity in doped polyacetylene. Phys. Rev. Lett. **39**(17), 1098–1101 (1977)
65. D. Moses, High quantum efficiency luminescence from a conducting polymer in solution—a novel polymer laser-dye. Appl. Phys. Lett. **60**(26), 3215–3216 (1992)
66. N. Tessler, G.J. Denton, R.H. Friend, Lasing from conjugated-polymer microcavities. Nature **382**(6593), 695–697 (1996)
67. F. Hide et al., Semiconducting polymers: a new class of solid-state laser materials. Science **273**(5283), 1833–1836 (1996)
68. I.D.W. Samuel, G.A. Turnbull, Organic semiconductor lasers. Chem. Rev. **107**(4), 1272–1295 (2007)
69. M.D. McGehee, A.J. Heeger, Semiconducting (conjugated) polymers as materials for solid-state lasers. Adv. Mater. **12**(22), 1655–1668 (2000)
70. B. Wenger et al., Mechanically tunable conjugated polymer distributed feedback lasers. Appl. Phys. Lett. **97**(19) (2010)
71. C. Gärtner, *Organic laser diodes: modelling and simulation* (Universitätsverlag Karlsruhe, Karlsruhe, 2009)
72. J. Clark, G. Lanzani, Organic photonics for communications. Nat. Photonics **4**(7), 438–446 (2010)
73. F.J. Duarte et al., Long-pulse narrow-linewidth dispersive solid-state dye-laser oscillator. Appl. Opt. **37**(18), 3987–3989 (1998)
74. H–.H. Fang et al., Distributed feedback lasers based on thiophene/phenylene co-oligomer single crystals. Adv. Funct. Mater. **22**(1), 33–38 (2012)
75. G. Wegmann et al., The dynamics of gain-narrowing in a ladder-type [pi]-conjugated polymer. Chem. Phys. Lett. **312**(5–6), 376 (1999)
76. T. Virgili et al., An ultrafast spectroscopy study of stimulated emission in poly(9,9-dioctylfluorene) films and microcavities. Appl. Phys. Lett. **74**(19), 2767 (1999)
77. L. Dal Negro et al., Applicability conditions and experimental analysis of the variable stripe length method for gain measurements. Opt. Commun. **229**(1–6), 337–348 (2004)
78. M.D. McGehee et al., Amplified spontaneous emission from photopumped films of a conjugated polymer. Phys. Rev. B **58**(11), 7035–7039 (1998)
79. L. Pavesi et al., Optical gain in silicon nanocrystals. Nature **408**(6811), 440–444 (2000)

80. Y. Tian et al., Spectrally narrowed edge emission from organic light-emitting diodes. Appl. Phys. Lett. **91**(14), 143504 (2007)
81. D. Amarasinghe et al., Broadband solid state optical amplifier based on a semiconducting polymer. Appl. Phys. Lett. **89**(20), 201119-3 (2006)
82. H. Goudket et al., Importance of dye host on absorption, propagation losses, and amplified spontaneous emission for dye-doped polymer thin films. Appl. Opt. **45**(29), 7736–7741 (2006)
83. C. Karnutsch, *Low Threshold Organic Thin-Film Laser Devices* (KIT, Karlsruhe, 2007)
84. T. Susdorf et al., Photophysical characterisation of some dipyrromethene dyes in ethyl acetate and covalently bound to poly(methyl methacrylate). Chem. Phys. **312**(1–3), 151–158 (2005)
85. D. Titterton, Solid-state dye lasers, in *Handbook of Laser Technology and Applications*, (three-volume set) (Taylor & Francis, Bristol, 2003), pp. 1143–1161
86. C.J. Lawrence, The mechanics of spin coating of polymer films. Phys. Fluids **31**(10), 2786–2795 (1988)
87. D. Meyerhofer, Characteristics of resist films produced by spinning. J. Appl. Phys. **49**(7), 3993–3997 (1978)
88. B. Park, M.-y. Han, Photovoltaic characteristics of polymer solar cells fabricated by pre-metered coating process. Opt. Express **17**(16), 13830–13840 (2009)
89. S. Klinkhammer et al., Continuously tunable solution-processed organic semiconductor DFB lasers pumped by laser diode. Opt. Express **20**(6), 6357–6364 (2012)
90. M.D. Rahn et al., Photostability enhancement of Pyrromethene 567 and Perylene Orange in oxygen-free liquid and solid dye lasers. Appl. Opt. **36**(24), 5862–5871 (1997)
91. S. Richardsonet al., Improved operational lifetime of semiconducting polymer lasers by encapsulation. Appl. Phys. Lett. **91**(26) (2007)
92. V. Navarro-Fuster et al., Highly photostable organic distributed feedback laser emitting at 573 nm. Appl. Phys. Lett. **97**(17), 171104 (2010)
93. E. Yariv et al., Efficiency and photostability of dye-doped solid-state lasers in different hosts. Opt. Mater. **16**(1–2), 29–38 (2001)
94. Y. Yang et al., Laser properties and photostabilities of laser dyes doped in ORMOSILs. Opt. Mater. **24**(4), 621–628 (2004)
95. A. Costela et al., Efficient and stable dye laser action from modified dipyrromethene BF2 complexes. Appl. Phys. Lett. **79**(3), 305–307 (2001)
96. R. Bornemann, U. Lemmer, E. Thiel, Continuous-wave solid-state dye laser. Opt. Lett. **31**(11), 1669 (2006)

Chapter 4
Organic Lasers Resonators

Abstract In this chapter we present the main resonators used for organic solid-state lasers, with a special emphasis on those based on thin-films such as planar waveguides or vertical external surface-emitting cavities. The influence of the resonator type on the laser properties (threshold, efficiency slope, output power, beam quality) is reviewed, with a special emphasis on works published after the year 2000.

The positive feedback loop induced by the resonator is one of the basic elements of any laser system. The presence of the resonator imposes two properties to the oscillating laser field. First, it defines the allowed resonant frequencies of the laser within the gain spectrum. These longitudinal modes correspond to the cavity resonances imposed by the phase conservation after a roundtrip in the resonator. Secondly, the spatial characteristics of the output beam are also a consequence of the filtering role of the cavity: only the modes that are self-replicating after a round trip can be amplified in the structure and ultimately define the transverse pattern of the laser beam. In every laser, the oscillating condition is that the gain has to overcome the various losses along a roundtrip in the cavity. The choice of the resonator parameters, and especially the control of the losses in the oscillating structure, is then of foremost importance to achieve lasing.

4.1 Cavities Made with Bulk Rods

Laser resonators may have various shapes and architectures, often more complex than the archetypal linear Fabry-Perot cavity or ring cavity, which are more suited to bulk laser gain media. Seminal works on organic lasers were however done with liquid dyes in such macroscopic cavities [1, 2]. This kind of resonators is composed of two mirrors placed around the polymer block or the cuvette containing the organic solution (see Fig. 1.1). One of the mirrors is often a diffraction grating: upon simple rotation, it is possible to select which wavelength will be

S. Forget and S. Chénais, *Organic Solid-State Lasers*,
Springer Series in Optical Sciences 175, DOI: 10.1007/978-3-642-36705-2_4,

retroreflected into the cavity to tune the laser. These macroscopic cavities usually allow very high output energies—up to the mJ level—usually associated with high thresholds [3]. With the same kind of macroscopic architecture used for liquid dye or crystalline solid-state lasers, it is possible to build external-cavity-based devices where a bulk organic gain medium is inserted inside a few-mm/cm-long cavity. These devices are appealing as they provide very efficient output energy extraction possibility due to an optimum overlap between the pump beam and the cavity mode. In addition, they have excellent spatial beam quality and have the potential of being tunable power-scalable devices.

In this respect, in 1993 Hermes et al. [4] reported on a pulsed, organic external resonator device with a slope efficiency as high as 85 % and an output energy of 128 mJ with a lasing threshold of 12 mJ/cm^2. In that experiment, the gain medium was a bulk plastic (hydroxypropyl acrylate/methyl methacrylate) cylindrical rod doped with Pyrromethene 570, placed inside a cavity with flat dichroic mirrors separated by 7.2 cm and end-pumped with a frequency-doubled Nd: YAG laser. Despite that excellent slope efficiency, the output beam was highly divergent as the authors measured a beam quality (M^2) value of 100. They attributed this high beam divergence to the use of a flat/flat (unstable) resonator combined with the effect of a 2 m positive focal length lensing in the rod itself. Following that success on the high laser efficiency, it was in 1997 that Faloss et al. [5] reported on a 10-cm-long plano-concave resonator enclosing a Pyrromethene-doped cylindrical rod of solid gel (Xerogel) obtained within a process called, sol-gel technique. This technique involves phase transformation of a "sol" obtained from metallic alkoxides or organometallic precursors. The sol, which is a solution containing particles (in our discussion an organic laser dye, however, this method is not limited to organic materials) in suspension, is polymerized at low temperature to form a wet gel. This one is then densified through thermal annealing to give an inorganic product like a glass or a dry gel. Using the gain medium obtained in this way incorporated in the above-mentioned architecture, Faloss et al. could achieve a nearly diffraction-limited emission (M^2 = 1.2) with a slope efficiency value of 86 %. In addition the authors could tune the emission wavelength of their device by introducing a prism inside the cavity leading to a tuning range extended over tens of nm. The laser was highly photostable and showed an excellent lifetime over hundreds of thousands of pulses when excited at 20 Hz repetition rate with pump energy of 1 mJ per pulse. The lifetime could further be improved to millions of pulses in oxygen-free Pyrromethene-doped samples.

All of the above-mentioned lasers are based on bulk organic gain blocks. For this kind of lasers, the fabrication of the gain block is a multi-stage process which sometimes can take up to few weeks. For example one of the fabrication techniques commonly used is the free-radical bulk polymerization technique [6]. The dye material and a free radical initiator are added to a monomer and the resulting mixture is agitated in an ultrasonic bath until complete dissolution of the dye is achieved. Then the resulting solution is filtered into appropriate cylindrical mold having an appropriate pore size filter. At this point the polymerization process starts upon careful increase of the temperature in discrete steps during several days. Then the temperature has to be reduced slowly to room temperature during several days until the sample is ready to be unmolded. The sample is not still ready to be used until it is cut parallel to the axis of the cylinder and the surface ends are delicately polished to optical-grade quality. Some other methods reported in the literature are sol-gel [7], hot-press molding technique [8] and use of epoxy matrices [9] that may also be complex in their own kind. Perhaps, this has been one of the reasons which made people think of realizing an "external-cavity" laser keeping all the advantages associated with this class of devices but with a simplified fabrication process. On this regard, one example is the nice work reported by Bornemann et al. on a continuous-wave solid-state dye laser [10]. Their setup (see Fig. 4.1) is a folded cavity comprising two concave mirrors and an output coupler together with a 50–100-µm-thick film of Rhodamine 6G-doped polymer sandwiched between two commercial DVD substrates as the gain medium and a birefringent filter for wavelength tuning. The gain medium was obtained through drop casting of a dye-doped UV-photosensible resist solution which was then cured by UV radiation. The disk-shape gain medium was mounted on a combined rotating/shifting translation stage at the Brewster angle and pumped longitudinally with a vertically-polarized diode-pumped solid-state laser. The continuous-wave operation was achieved thanks to the disc rotation since exciting the organic dyes in continuous regime fills the long-lived triplet state that is detrimental to laser operation and eventually extinguishes the laser emission (see details in Sect. 2.6) Using this technique, they could achieve a slope efficiency of 2 % and a laser emission that could last for 30 min but with strong fluctuations (300 %) attributed to mechanical imperfections and spatial inhomogeneities in the gain layer thickness. The wavelength tuning range was extended over 40 nm. Recently, the same team proposed an improved setup with more stable operation and higher output power [11]. The strategy adopted in this work to achieve continuous operation and avoid fast photodegration (mimicking the dye jet flow behavior in liquid dye lasers) can be considered as a milestone in the development of organic solid-state lasers.

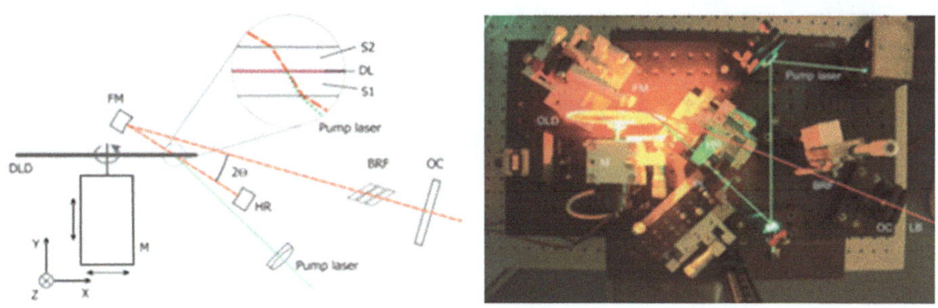

Fig. 4.1 (*Left*) scheme of the CW laser based on a spinning disc developed in [10, 11]. (*Right*) photograph of the system. Courtesy of R. Bornemann [12]

4.2 Cavities Based on Thin-Films

As highlighted in the previous chapter, a major advantage of organic semiconductors and dyes embedded into polymeric matrices is their ability to be easily processed under the form of thin films of micrometric thicknesses or less, through low-cost, sizeable and controllable fabrication techniques. Small-molecular organic semiconductors may additionally be thermally evaporated, a technique which provides an excellent control over thickness and optical quality. The most natural resonator geometry for thin-film organic lasers is consequently the two-dimensional planar waveguide. We will focus in the following paragraphs on those "thin film" lasers, and give a rapid review of some of the most frequently used resonator geometries.

Organic materials can be easily and naturally deposited under the form of thin films on various substrates, such as microcavities, planar waveguides, periodically corrugated structures, microdisks or microrings as illustrated on Fig. 4.2. Lasing is also observed in organic thin films under intense pumping in the absence of cavity because of scattering (either intentionally added with scattering nanoparticles, or just naturally present because of film inhomogeneities): this is referred to as random lasing.

The shape and the thickness of the film can be controlled in a relatively simple way through evaporation or spin coating techniques, giving birth to a wide panel of resonators. We will now review some of them.

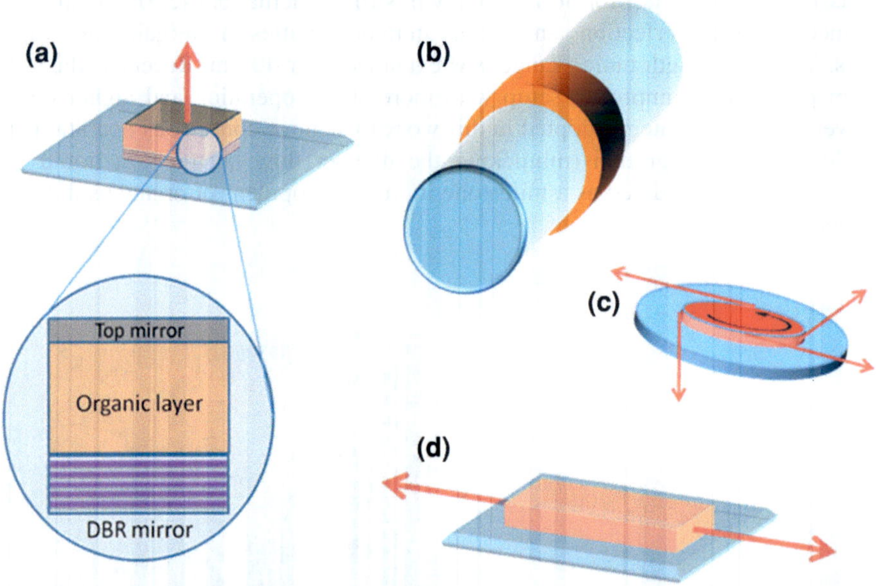

Fig. 4.2 Examples of resonators used for organic semiconductor lasers: **a** vertical Fabry-Perot microcavity, **b** microring resonator, **c** microdisk resonator, **d** Planar Fabry-Perot (or DFB/DBR, see Figs. 4.5 and 4.6) waveguide

4.2.1 Waveguide Lasers

In a planar waveguide laser, the resonator axis is parallel to the film plane: photons travel a long way (several millimetres) through the active medium during a roundtrip, leading to high overall optical gain. In this geometry, light is waveguided in the high refractive index organic layer, sandwiched between, typically, a low-index substrate (glass or silica) and air. As the typical order of magnitude for an organic layer thickness is around 1 μm, single-mode waveguides are easily achieved. Optical feedback can be obtained in several ways reviewed below.

A waveguide is a structure that confines light in a medium by total internal reflection at the medium interfaces with the surrounding media. For light to be trapped and guided, the refractive index of the propagating medium must be higher than the index of the surrounding media. The amount of light confinement is determined by the refractive index contrast between the propagating medium and the surrounding media. The bigger the difference, the more the light is confined and guided within the structure. One of the most encountered waveguide structures is the planar (slab) geometry where a thin film of a material with refractive index of n_{film} is sandwiched between the up and down claddings. If the claddings have the same refractive indices, then the structure is a symmetric planar waveguide. A very simple case that we are dealing with in this context is an asymmetric waveguide where the lower cladding is made of fused silica (SiO_2) with $n_{sub} = 1.45$ and the upper cladding is the air with the refractive index of $n_{cladding} = n_{air} = 1$.

A schematic representation of an asymmetric slab waveguide structure is shown in Fig. 4.3.

The propagation of light in the above-illustrated structure composed of three isotropic and homogeneous media, labeled with $i = 1, 2, 3$ each one with a dielectric constant $\varepsilon_{r,i} = n_i^2$ can be described by the Helmholtz equations for the electric and magnetic field components of light:

$$\begin{aligned} \nabla^2\vec{E} + k_0^2 n_i^2 \vec{E} &= \vec{0} \\ \nabla^2\vec{H} + k_0^2 n_i^2 \vec{H} &= \vec{0} \end{aligned}$$

where $k_0 = \frac{2\pi}{\lambda}$ is the wave vector, and **E**(r) and **H**(r) are the electric and magnetic fields, respectively. The solutions to the above equations for an electromagnetic field of light propagating along the z direction are given by:

$$\begin{aligned} \vec{E}(x,y,z,t) &= \vec{E}(x,y)e^{j(\omega t - \beta z)} \\ \vec{H}(x,y,z,t) &= \vec{H}(x,y)e^{j(\omega t - \beta z)} \end{aligned}$$

where β is the propagation constant defined as: $\beta = \frac{\omega}{c} n_{eff} = k_0 n_{eff}$; n_{eff} is the effective refractive index, that is, the refractive index seen by the waveguided light along the propagation direction and c is the speed of light. The n_{eff} value lies between the lowest and the highest refractive index, that is between the index of the substrate (or upper cladding) and the index of the gain material in the middle, supporting the guided mode.

Generally, in waveguide problems only the transverse electric (TE) modes are considered since the treatment of TM (transverse magnetic) modes is similar. In addition, the field variation along the x direction is assumed to be negligible as the waveguide is considered to be infinite in the direction.

Considering the above assumptions, the expansion of Helmholtz equations for the electric field propagating along the z direction will be, in each slab of index n_i:

$$\frac{\partial^2}{\partial y^2}\vec{E}(y) + k_0^2\left(n_i^2 - n_{eff}^2\right)\vec{E}(y) = \vec{0}$$

Solving this equation for each of the media within the waveguide structure (upper cladding, film and the substrate) with application of appropriate boundary conditions at the interface layers will determine the confined guided modes. These imposed conditions are the continuity of the transverse electric profile and its derivative at the layer boundaries.

It appears that the guided solution has a sinusoidal shape in the confining film and decays exponentially decay in the cladding layers.

The full description of the waveguide theory will not be given here, but it can be easily found in any waveguide textbook [13].

4.2.1.1 Fabry-Perot Waveguides

A convenient and low-cost arrangement used in inorganic semiconductor diode lasers to form the cavity mirrors is to cleave the semiconductor to obtain flat facets and form a plano–plano Fabry-perot waveguide cavity. In this geometry light is confined and guided inside a medium sandwiched between two up and bottom claddings having lower refractive indices than the medium, such as the one presented in Figs. 4.3 and 4.4. When guided light emerges from a cleaved facet, it experiences a brutal change in refractive index (generally the light is outcoupled in

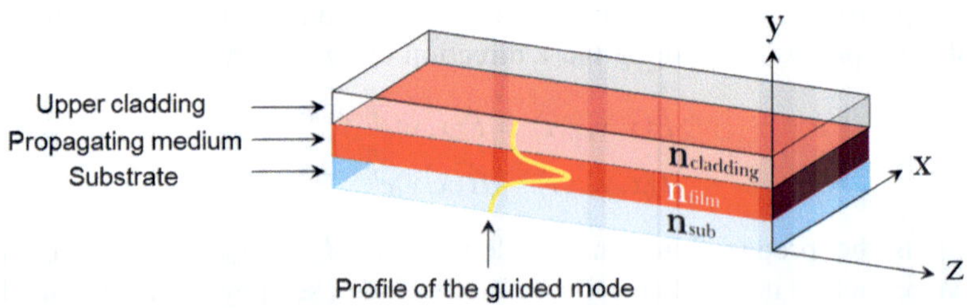

Fig. 4.3 Three dimensional representation of planar waveguide geometry. For light to be guided one should have $n_{film} > n_{sub}$ and $n_{film} > n_{cladding}$

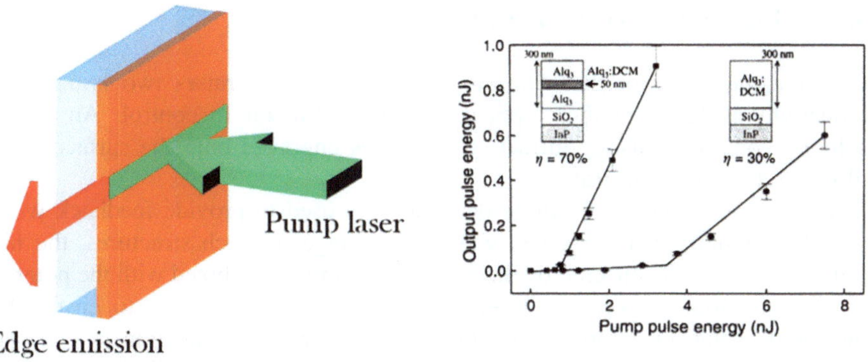

Fig. 4.4 Schematic representation of the waveguide structures realized by Kozlov et al. [15], together with the corresponding input–output characteristic curves. Courtesy of S.R. Forrest. Reprinted from [15] Copyright (1999), with permission from Elsevier

air) and hence a reflection. For an infinite medium of index n, the reflectivity at normal incidence between this medium and air is given by $R = \left(\frac{n-1}{n+1}\right)^2$. With inorganic semiconductors, the reflectivity is as high as ~30 % due to the very high refractive index of the active medium. With organic materials, as the refractive index is typically two times lower (n is typically ~1.7–1.8 for organic semiconductors, ~1.5 for dye-doped polymers), the "natural" reflectivity at the edges of waveguides does not exceed a few percents. Moreover, facets with optical quality are difficult to obtain in thin films due to the amorphous and soft nature of the materials. Despite all these difficulties, Kozlov et al. [14] reported on a Fabry-Perot laser structure (Fig. 4.4) based on a vacuum-deposited film of Alq_3:DCM (n = 1.7) with end facet mirror reflectivities of 7 % obtained naturally during evaporation process. In that experiment high-level losses imposed by low mirror reflectivities were compensated by choosing a rather long (~1 mm) Fabry-Perot waveguide. In such a structure, a quantum slope efficiency[1] of 30 % was obtained but the authors demonstrated that it could be improved to the remarkable value of 70 % (see Fig. 4.4) with considerably reduced lasing threshold of 1 μJ.cm^{-2} using a double-heterostructure device where a film of Alq_3:DCM was sandwiched between two undoped Alq_3 cladding layers. Despite excellent slope efficiency, the output energy was less than 1 nJ per pulse and the emission was suffering from low spatial quality (see Sect. 1.4.2) with high divergence due to sub-wavelength edge-emission aperture.

[1] The quantum slope efficiency η_q is a criterion for laser performance assessment and is determined by measuring the slope of the laser input–output characteristic curve expressed in terms of photons instead of energy. Expression of the slope efficiency in terms of energy is known as the energy slope efficiency (η_e) and is defined by $\eta_e = \left(h\nu_{laser}/h\nu_{pump}\right)\eta_q$.

4.2.1.2 Distributed Feedback Structures

Cleaving the facets of thin films to obtain laser effect raises two issues: their reflectivity is low and their optical quality is difficult to control. An elegant solution consists in using diffractive structures engraved onto the surface of the film to act as cavity mirrors.

In this case, a periodic diffractive grating is used to provide feedback with a very high efficiency for a given wavelength range. In such structures, the high reflection coefficient ensured by the periodic grating is combined with the potential long interaction between the laser wave and the gain medium to ensure low threshold lasing. One can classify the periodic surface corrugation in two categories, namely Distributed Bragg resonators (DBR) and Distributed FeedBack (DFB) structures.

A DFB laser consists of a thin active layer deposited onto a corrugated substrate. The light propagates in the waveguide and is scattered by the periodic corrugation. If the scattered waves combine coherently, a Bragg wave is created in a new direction. The full description of such a mechanism requires the use of coupled mode theory and has been described by Kogelnik et al. [16]. For a given period of modulation, only a given set of wavelengths will correspond to constructive interferences in the direction opposite to the incident wave and leads to the feedback required for laser operation (Fig. 4.5). The bandwidth (or width of the "stopband" around the central Bragg wavelength) is related to the modulation depth. The central Bragg wavelength is given by the Bragg condition:

$$m\lambda_{Bragg} = 2n_{eff}\Lambda$$

where λ_{Bragg} is the wavelength of the laser light, Λ is the period of the corrugation, and m is an integer representing the order of the diffraction process. n_{eff} is the effective refractive index of the waveguide. For first order diffraction process ($m = 1$), we obtain $\Lambda = \lambda/(2n_{eff})$.

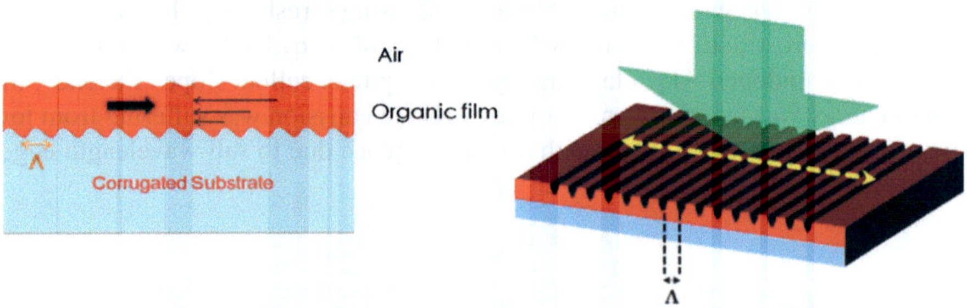

Fig. 4.5 Possible structure of a DFB laser where the organic film is deposited on a pre-corrugated substrate with a modulation period equal to $\Lambda = m\lambda/2n_{eff}$. The light propagating from the left is scattered by the corrugation, and the diffracted waves interfere constructively in the opposite direction, creating a counterpropagating wave

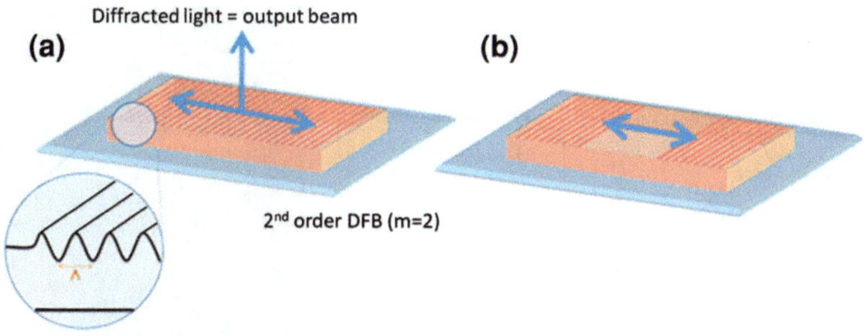

Fig. 4.6 **a)** DFB structure directly engraved onto a polymeric film. If $\Lambda = \lambda / n_{eff}$ (2nd order grating, m = 2) then a diffracted beam (first diffraction order) is emitted in the direction perpendicular to the film plane, **b)** DBR structure with no corrugation between two Bragg mirrors

An issue encountered with such a structure is light extraction through highly reflective Bragg mirrors, together with the high divergence due to the small transverse size of the waveguide. To circumvent both problems, the second order grating (m = 2) is of particular interest. Indeed, in this configuration the resonant wavelength is $\lambda = n_{eff}\Lambda$: the grating diffraction condition for this order imposes that some light is diffracted perpendicular to the waveguide plane, leading to surface emission [17, 18] (Fig. 4.6a). Such second-order structures can provide an efficient and practical way to extract light out of the cavity, even if second order gratings typically lead to a higher lasing threshold compared to first order gratings [19, 20].

One potential difficulty with a DFB structure is that the thickness modulation can induce incoherent scattering associated to the surface irregularities, and consequently a high level of losses. Lower threshold can thus be obtained through the use of so-called *Distributed Bragg Reflector* (DBR) geometries [21]. The term was usually applied to a stacking of quarter wavelength layers with alternating high and low indices (in dielectric mirrors for examples). A DBR laser consists of an organic waveguide in which two outer areas are nanopatterned to ensure Bragg reflection [22, 23] (see Fig. 4.6b). Each of the corrugated areas acts as a Bragg mirror, and reflects light for the chosen wavelength. The advantage of such a structure is that the gain medium located in-between remains uncorrugated, which limits losses induced by scattering.

Simple one-dimensional DFB or DBR organic lasers have been reported in many materials (polymers [17, 18, 24, 25] and small molecules [26–31]) covering almost all the visible spectrum.

An advantage of DFB/DBR lasers is that they emit a nice monochromatic light as the Bragg condition on the wavelength is restrictive, the spectral linewidth of DFB or DBR lasers is typically well below 1 nm. Moreover, this wavelength can be relatively easily tuned by changing either the effective index (which can be done through a change of the thickness of the organic layer for instance) or the period of the modulation (see Sect. 6.1.3). The concept of DFB laser can also be extended to two-dimensional systems. This type of periodic structures (where

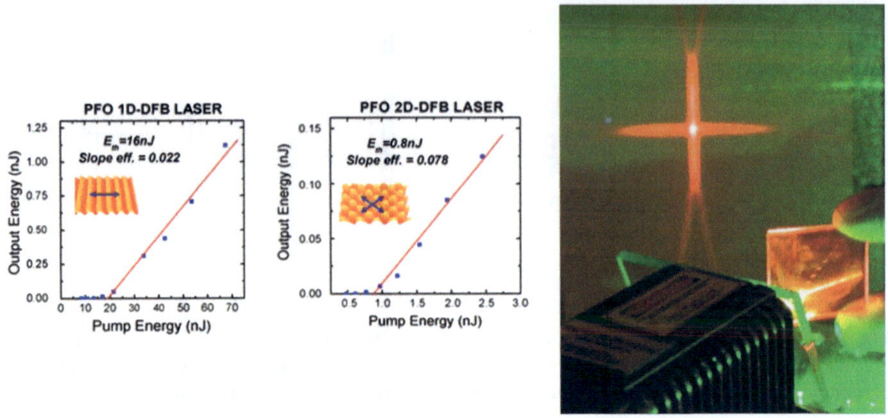

Fig. 4.7 Improvement on the lasing threshold provided by a 2D corrugation (courtesy of G. Turnbull, adapted from [36]) on a polyfluorene laser. *Right* Example of far-field structure of a diffracted laser beam (courtesy of I. Samuel and G. Turnbull, reprinted from reference [37] in Sect. 6 of this book, with permission from Elsevier, copyright 2004)

either the thickness or the refractive index is modulated in 2D) is referred to as 2D photonic crystal structures [32, 33]. There is a growing interest in such structures, either with square, hexagonal or circular concentric geometries. In such resonators, optical feedback is applied in several directions in the plane of the film. This opens the way toward photonic bandgap engineering where the spectral and spatial characteristics of the laser beam could be controlled with a high degree of accuracy. For example, it has been demonstrated by Ifor Samuel's group at St Andrews (see Fig. 4.7) that a clear enhancement of the spatial beam quality can be obtained through the use of 2D DFB structures [32–35] (even if the far-field pattern generally reflects the grating geometry, for example forming a cross in a square-lattice photonic crystal [36]) together with an important decrease of the threshold (for example by a factor of 20 compared to a standard 1D DFB laser [35]).

Another improvement has been demonstrated through the use of mixed first and second order DFB gratings. The first order regions (grating period Λ_1) ensure very efficient feedback in the cavity. The second order region (grating period $\Lambda_2 = 2\,\Lambda_1$) placed in-between the first-order regions, outcouples light perpendicularly to the waveguide plane. With such a configuration a threshold as low as 36 nJ/cm^2 has been demonstrated with a F8BT polymer emitting at 452 nm [37].

4.2.1.3 Random Lasers

The high gain available in organic thin films means that it is quite easy, with available pulsed lasers as pump sources, to obtain a laser effect. Even without a cavity, as seen in Sect. 3.2.1, Amplified Spontaneous Emission can be observed, characterized by a narrow spectrum (a few nm FWHM), a threshold (when gain surpasses the passives losses of the material) and a superlinear dependence of ASE

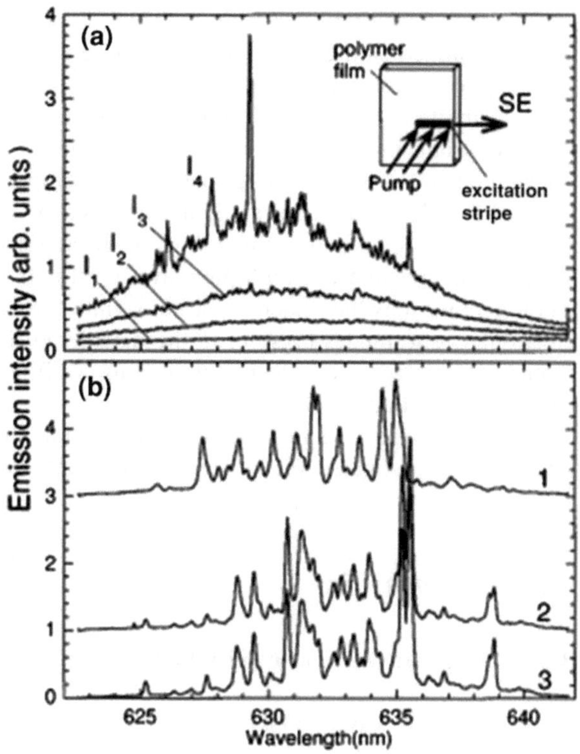

Fig. 4.8 (**a**) Stimulated emission spectra observed in DOO-PPV film obtained using a stripe-like excitation area (length L = 1 mm, width a = 30 μm, at different excitation intensities (I1 = 1 MW/cm^2, I2 = 1.25 I1, I3 = 1.4 I1, I4 = 1.6 I1). (**b**) SE spectra measured sequentially above random lasing threshold; 1 and 2 correspond to different pump spot locations: 2 and 3 correspond to the same spot location but at 3 min time interval : the spectra are similar. Reprinted with permission from [38]. Copyright (1999) by the American Physical Society

intensity with the size of the pumped area on the film. This kind of emission, sometimes referred to as "mirrorless lasing", is however not "laser" emission, in the sense where there is no feedback and no modes in the emission.

However, for pump intensities which are usually substantially higher than the ASE threshold, a strong structuration of the ASE spectrum may be observed, some very sharp peaks (~0.1 nm width) appearing after a clear threshold (see Fig. 4.8) [38]. These sharp peaks have a spectral position and a relative intensity which is very much dependent on the pump location onto the sample, but usually do not vary with time from one pulse to another, except if strong competition exists between these "modes", which may happen near threshold [39]. These unusual modes are characteristics of *random lasing*, a general lasing operation observed in highly-scattering gain media [40, 41].

Random lasing in organic lasers was first observed in liquid dye solutions in which TiO_2 nanoparticles were added to act as scatterers [42]. Random lasing was then observed in organic solid-state thin film lasers even without any intentionally added scatterer, in pi-conjugated polymers [43] or in small-molecular neat films [44]. In organic films, scattering is supposed to be brought by defects present in the film (porosity, dust onto the surface, aggregates, etc.) which induce the scattering required for random lasing. In a theoretical paper, Vanneste et al. [45] have shown that random lasing with coherent feedback is possible even in weakly-scattering media inasmuch as the pumping rate is large enough.

The physical origin of random lasing has been a subject of long controversies [40]. A simple and appealing picture consists in assuming the existence of closed loops for scattering forming random ring cavities inside the medium. However this image, consistent with a strong localization comparable to Anderson localization [46–49] for electrons in solids, is not necessarily adequate for weakly scattering media such as organic thin films. Instead, if we remind that in a traditional laser, sharp modes result from multiple-wave constructive interferences inside a resonator, here we may have such constructive interference between the many complex electromagnetic modes existing in the disordered medium: similar random interference effects are seen when illuminating a diffusive plate by a coherent light beam and is called speckle.

Random lasing is a fascinating fundamental phenomenon that has not been intensively investigated yet in organic solid-state lasers.

4.2.2 Microcavities

4.2.2.1 Organic Vertical Cavity Surface-Emitting Laser

Another natural way of making an organic laser, inspired by stacked multilayer organic light emitting diodes, consists in a planar sandwich of the organic material between two mirrors to form a Fabry-Perot microcavity in the vertical direction (resonator axis is perpendicular to the film plane, see Fig. 4.2a). This kind of configuration has already been successfully used for inorganic diode lasers where it is known as Vertical Cavity Surface Emitting Lasers (VCSELs). The Organic Vertical Cavity Surface-Emitting Laser (OVCSEL) geometry was the one used by Tessler et al. [50] to demonstrate optically-pumped organic lasing in PPV for the first time. Using a highly reflective broadband DBR mirror and a partially transmitting silver mirror as an output coupler, a high quality factor Q could be achieved [51]. The 100-nm-thick PPV layer was simply spin coated onto the DBR mirror and the silver layer was evaporated on top. Because the optical length of a microcavity is of the same order of magnitude as the optical wavelength, only a few longitudinal modes are supported by a microcavity, and when the pump power exceeds the lasing threshold (200 $\mu J/cm^2$ in this case), a single mode dominates the emission spectrum of the laser. Following this pioneering work, several planar microcavities with various organic media were reported. The microcavity Q-factor (and thus the lasing threshold) can be improved by sandwiching the polymer between two DBR mirror [52], or by directly growing the top DBR mirror at low temperature onto the polymer [53]. Another option is to change the lasing medium to use small molecules instead of polymers. Lasing with Alq_3:DCM blend in a DBR/metal [51] or DBR/DBR [54] microcavities were reported. Recently, Sakata et al. [55] realized such a microcavity laser with a cheap and compact UV laser diode as a pump source.

The problem linked to the poor beam quality of diffractive DFB/DBR lasers is much more improved with Organic Vertical Cavity Surface Emitting Lasers (OVCSEL), which exhibit a circular well-defined and measurable beam. However the output beam of an OVCSEL is not, a priori, diffraction-limited (see definition of diffraction limit and M^2 factor in Sect. 1.4.2). This is because the plano–plano cavity formed by the two plane mirrors is not optically stable, i.e. a given distribution of light inside this structure cannot be self-replicating after one roundtrip, except if some stabilizing phenomenon (gain guiding, index guiding, thermal lensing) exists.

The threshold in these devices is normally much higher in comparison with DFB devices since the gain medium is extremely thin and thus a very high gain per unit length is required [56]. In terms of tunability, in a recent work Schütte et al. [57] has reported on a continuously tunable laser emission from a wedge-shaped organic microcavity where, depending on the position of the excitation beam over the varying film thickness, a tuning range extended from 595 to 650 nm could be obtained.

In terms of laser efficiency, similarly to diffractive resonators, organic VCSELs have efficiencies which do not exceed a few percents.

4.2.2.2 Microdisks, Microrings and Microspheres

The easy processing of organic materials enables the realization of new geometries that are impossible or much more difficult to realize with classical inorganic semiconductors.

A first example is the microring laser, where a polymeric film is deposited around an optical fiber or even a metallic wire [58]. This can simply be achieved by dipping the fibre into a solution of the polymer [59]. The waveguide created around the fibre forms a ring microcavity where the light bounces through total internal reflection on the interface between the polymer and the surrounding medium. As the diameter of the fibre is typically around several micrometers, the light travels through a "long" (several wavelengths) path to make a roundtrip, which is associated to a complex mode pattern, mixing whispering gallery modes and waveguide modes [60]. As the losses for these modes can be made very low, very low thresholds can be achieved with such structures [61] (around 1 $\mu J/cm^2$). The pumping of these lasers can be done transversally or through the core of the fibre, leading to a more uniform pumping pattern [62].

In a similar way, microdisks lasers [61, 63] can be realized by lithographically patterning and etching an organic film to form circular disks of several

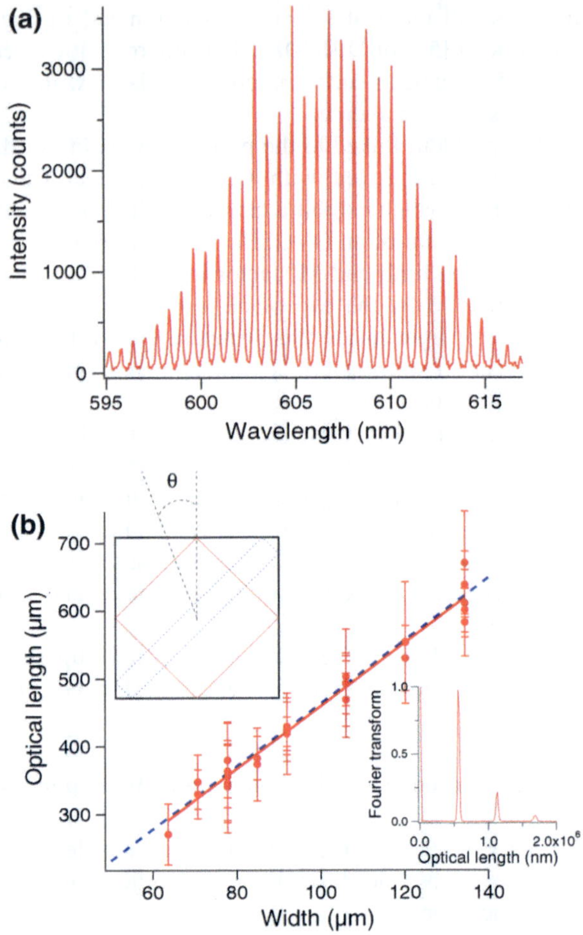

Fig. 4.9 (**a**) Optical spectrum obtained from a square-shaped PMMA:DCM microlaser (square width W = 150 μm), from ref [65]. (**b**) A Fourier transform (inset) of the spectrum allows inferring directly the length of the periodic orbit, which as expected is found to vary linearly with the square dimension W. Courtesy of M. Lebental. Reprinted with permission from [65]. Copyright (2007) by the American Physical Society

micrometers in diameter. The microdisk supports whispering gallery modes comparable to those observed in microring lasers, but also possible Fabry-Perot resonances (when light is bouncing back and forth from opposite parallel parts of the disk). With more complex shapes [64] like squares, pentagons, stadiums… a wide variety of mode patterns are addressable, which offers a tool to study quantum chaos or chaotic lasing. The regular comb-like spectrum obtained even with complex shapes enables inferring the dominant periodic orbits in such structures [65] (see Fig. 4.9).

Finally, the microsphere structure [66] is the 3D generalization of the microdisk. It can be formed by superimposition and melting of several microdisks on a lyophobic surface to form droplets. A common property shared by these geometries is the messy output, since light is emitted in many radial directions. This unusual output may be tailored upon playing on the shape of the resonator [64].

4.2.3 Vertical External Cavity Surface-Emitting Organic Lasers

A major drawback shared by DBR/DFB, microdisks/rings and microcavity organic lasers is the asymmetric and/or highly diverging profile of the output beam, which limits the potential applications of such laser sources. *A contrario*, circular and diffraction-limited beams (i.e. with the minimum beam divergence obtainable from a beam of given transverse size, governed by diffraction laws) can be obtained with external bulk resonators using dye-doped polymer blocks but at the expense of compactness and with longer and more complex fabrication processes than organic thin films.

Hence, combining the advantages of thin films (ease of fabrication, low cost, etc.) with that of external resonators (efficiency, beam quality, good control of laser emission through the insertion of intracavity elements) upon building a *thin-film* based organic laser in an *external cavity* is an attractive goal.

A thin-film-based external-cavity device has been reported by Zavelani-Rossi et al. [67]. In this work, a stable plano-concave resonator, containing a 500-nm-thick neat film of electroluminescent oligothiophene, deposited directly onto the high reflector via the fast and simple spin coating technique, was shown. The whole structure was end-pumped with 150-fs-long pulses from the second harmonic of Ti: sapphire laser. Using this configuration, they reported on a nearly diffraction-limited beam with 1.7 % efficiency and a tunability range that was achieved over 30 nm by varying the resonator length (because the laser was single-mode), which could be increased up to ~ 6 μm. No lasing was observed for cavity lengths exceeding this value. This short cavity length limit can be due to the low available gain of the material and the pump length duration which govern the laser pulse buildup time. Considering the modest efficiency value together with the short maximum cavity length obtained, the setup is very much similar to a microcavity device. Before this work, in 1998, Schülzgen et al. [52] reported on a similar structure based on a conjugated polymer where the cavity length was also limited to 9 μm.

In these two examples the cavity is too short to enable the insertion of intracavity elements to control the emission, for instance frequency converters or polarizers; and the conversion efficiency (laser output energy/pump pulse energy) is lower than what can be expected from such resonators.

The key to these issues is the choice of the pulse width of the pump laser and the optimum output coupling, depending on the available gain in the thin film. This was shown in a recently demonstrated laser resonator called Vertical External Cavity Surface-emitting Organic Laser (VECSOL, see Fig. 4.10) [68] which is the organic counterpart of the inorganic VECSEL [69]. It is composed of a plane mirror coated with a thin film of the organic material and a remote concave mirror to close the cavity (see Fig. 4.9). The macroscopic (up to several centimeters)

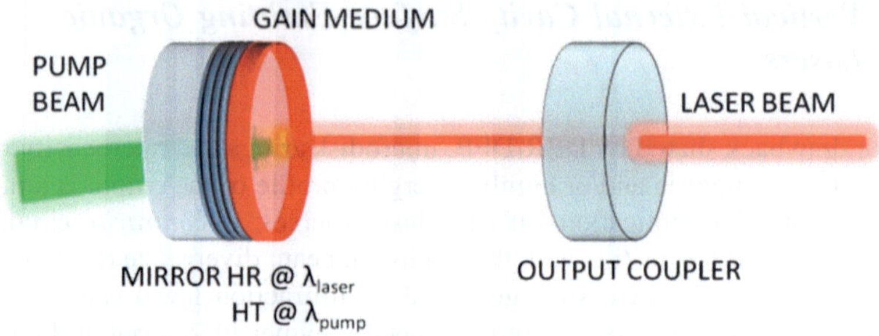

Fig. 4.10 A vertical external cavity surface-emitting organic laser (VECSOL) [68]. The left mirror is transparent for the pump wavelength and highly reflective for the laser radiation. The output coupler radius-of-curvature and reflectivity are optimized to yield the highest output energy in a single transverse mode

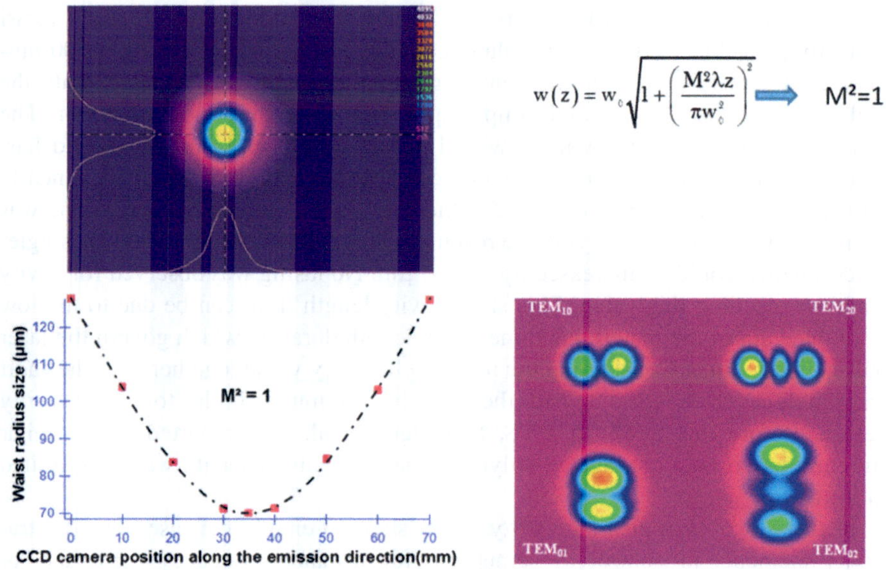

Fig. 4.11 Spatial characterization of the beam emitted by a VECSOL. The M^2 is equal to 1 (diffraction limited beam) with a perfectly Gaussian beam. Slight misalignment of the cavity leads to the observation of the various higher order modes predicted by laser resonator theory

cavity defines the spatial geometry of the laser mode and gives birth to a TEM_{00} diffraction-limited transverse profile (see Fig. 4.11). This laser architecture also leads to one of the highest optical conversion efficiency reported for thin-film organic lasers (more than 50 % [70], see Fig. 4.12). This contrasts with the relatively low slope efficiencies obtained with DFB structures (usually a few percents or less).

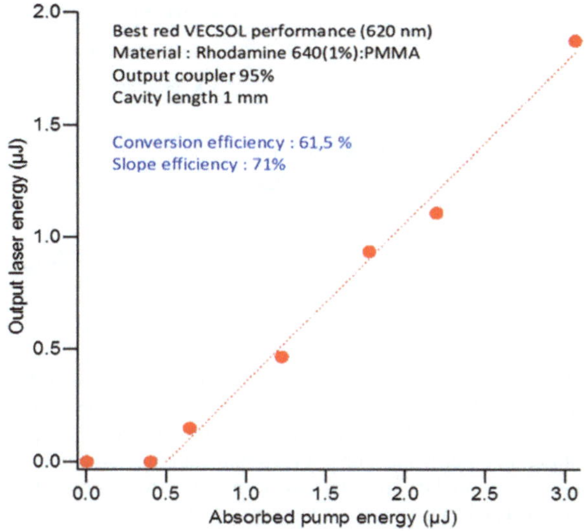

Fig. 4.12 Output pulse energy versus absorbed pump energy in a VECSOL, showing the very high. Here the gain thin film is a 17-μm-thick PMMA layer doped with 1 wt% of rhodamine 640, spin-coated onto the rear mirror

Another interesting feature of the VECSOL is *power scalability* or in other words the possibility to obtain high energies with a constant beam quality and a constant photodegradation rate, upon increasing the pump beam size on the sample (in a laser structure where the modes have a given spatial extent, such as a DFB laser or a microdisk, the only way to obtain higher output energies consists in increasing the pump energy density, which goes with a faster degradation). Power scalability in a VECSOL is illustrated in Fig. 4.13.

The study of the turn-on behavior and pulse creation dynamics in a VECSOL structure provide interesting insights into the physics of this kind of lasers. Usually in thin-film compact resonators, the cavity photon lifetime is short compared to the gain window (the time during which gain is present in the material, governed approximately by the highest time between the pump pulse duration and the fluorescence lifetime), meaning that the laser pulse shape will typically follow the pump pulse and will have similar duration. In VECSOLs, pulse buildup dynamics are responsible for the large dependence of performance on the cavity length, and are therefore of high relevance.

The efficiency of an organic VECSEL depends naturally on well-known parameters (pump intensity, output coupling, active material spectroscopic parameters and concentration, etc.). In the case of organic VECSELs, which need to be operated in a pulse mode (see Sects. 2.5 and 2.6), the efficiency depends also critically on the pulse buildup time, which depends on the pump pulse duration, pump intensity and cavity length [70].

Indeed, as an organic VECSEL is the association of an external cavity with a thin-film active medium, a laser photon spends most of the round-trip time in passive areas (mostly air) rather than in active regions, which makes it clear that

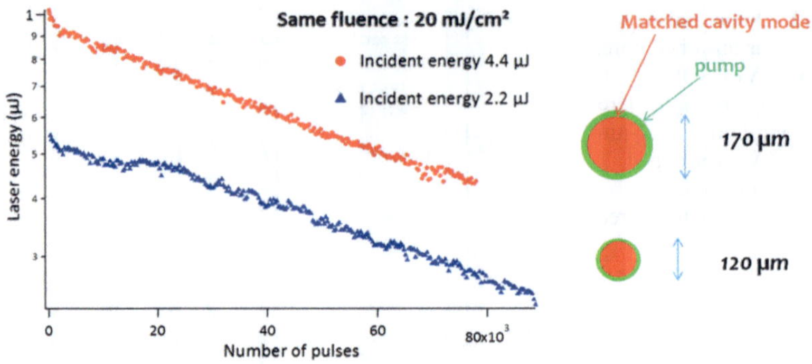

Fig. 4.13 Illustration of the power scaling capability of VECSOLs. The output coupler has a 98 % reflectivity with 50 mm radius of curvature. The TEM_{00} fundamental mode waist diameter corresponding to this curvature is 104 µm. Therefore, a pump waist diameter of 120 µm was chosen in order to have only a predominant pump area over the TEM_{00} fundamental cavity mode and not the higher order transverse modes. This corresponds to a pump energy of 2.2 µJ (*blue curve*). Then a pump waist diameter of 170 µm was chosen in order to enlarge the pump area by a factor of two. Meanwhile, the pump energy was doubled to 4.4 µJ to ensure that the energy density was still the same at 20 mJ/cm^2 ($\sim$two times above threshold). The output coupler was also replaced by an output coupler of 98 % reflectivity and 200 mm radius of curvature to match this pump size and the fundamental TEM_{00} cavity mode (having in this case 150 µm waist size). The experimental results reveal that the output laser energy is almost doubled in the second case, as expected, but they also reveal that the degradation rates are equal in both cases

the highest efficiency will be obtained for short cavities, long/intense pump pulses, all of which help minimizing the pulse buildup time and make it fit within the short time window during which gain is available. This limitation is most of the time ignored in inorganic VECSELS which are operated in the CW regime as it only plays on the turn-on dynamics and not on the efficiency. In VECSOLs we can expect that the laser pulse will be broadened and delayed when the laser is operated with long cavities, and also that the efficiency will drop significantly when the cavity length is increased.

This effect of broadening and delaying can be verified experimentally as illustrated in Figs. 4.14, and 4.15. The experimental conditions are summarized in Fig. 4.14. A "short-pulse" pump source is used with pulse widths of about 0.35 ns at 532 nm.

The pulse shapes recorded for $L = 0.3$ mm (corresponding to the 1″-in diameter mirrors touching at the apex), $L = 1$, 5 and 10 mm are shown in Fig. 4.15 (all traces have been normalized). The tendency of broadening and shifting is well taken into account. The shift is especially visible for the 10-mm-long cavity, where the laser pulse is clearly temporally separated from the pump pulse. These results can be very well accounted by a simple numerical model presented in [70].

The preceding results suggest that a short cavity and a long-pulse pump source are hence desired to obtain the highest efficiency with a VECSOL: in this case indeed, we can make sure that enough time is available to reach steady-state, i.e. to

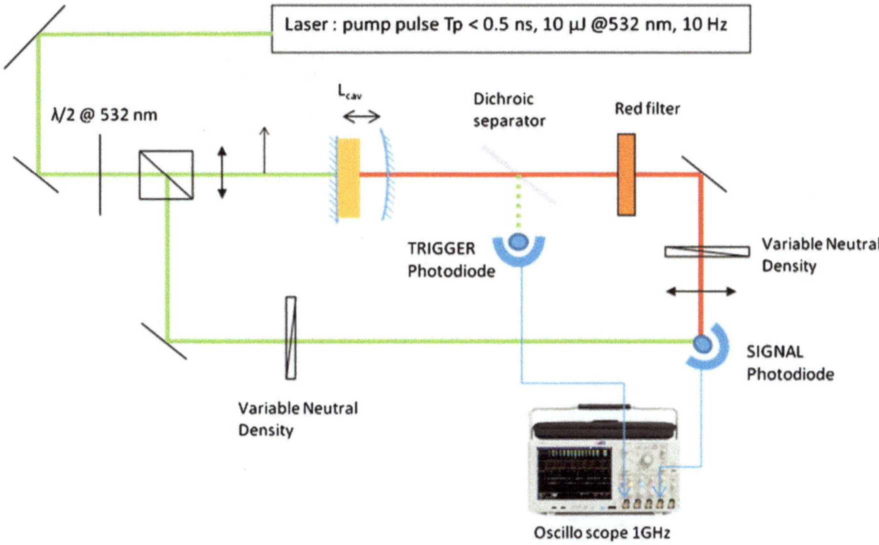

Fig. 4.14 Experimental setup for simultaneous measurement of pump and laser pulse shapes, for different laser cavity lengths in a VECSOL. The pulse shapes are measured with a fast photodiode connected to a 1-GHz oscilloscope. In order to precisely measure the shifts of the laser pulse with respect to the pump pulse, both pump and laser pulses are measured with a unique photodiode, set up in such a way that there is no optical delay between the pump path and the laser path. The trigger for the oscilloscope is provided by a photodiode capturing the pump radiation exiting from the dichroic separator used to remove the unabsorbed pump radiation after the laser cavity

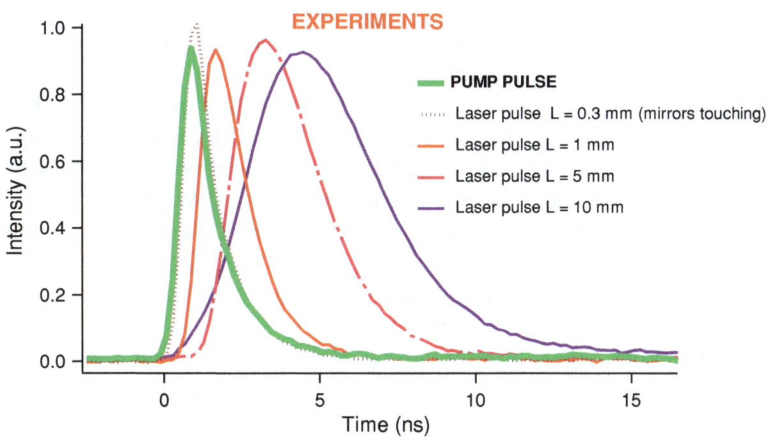

Fig. 4.15 Measured pulse shapes of pump (*green line*) and laser beams for different cavity lengths (L = 0.3, 1, 5 and 10 mm) in a VECSOL pumped by 0.35-ns pump pulses. The active layer is 17-µm thick. The signals are recorded with the setup shown in Fig. 4.14

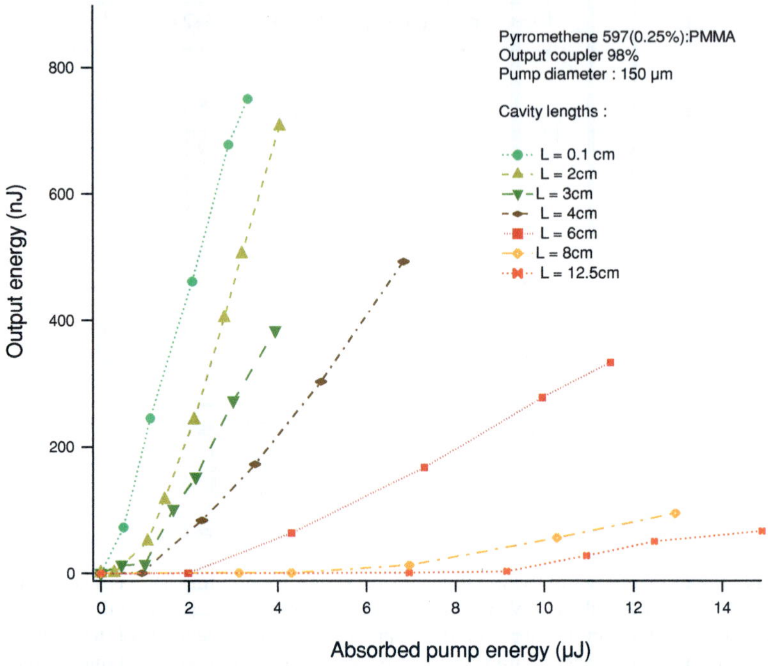

Fig. 4.16 Network of efficiency curves measured for different cavity lengths for a rhodamine 640:PMMA VECSOL pumped by a 25-ns pulsewidth laser. It is observed that when long cavities are used, the slope efficiency if lower and the threshold is higher

let the inversion clamp to its threshold value and convert pump photons into laser photons with the highest efficiency possible.

In Fig. 4.16 are presented the efficiency curves of a given VECSOL with a highly photostable material, Pyrromethene 597 in PMMA [71]. The pump source is a diode-pumped Nd:YAG laser (Harrier from Quantronix Inc.) followed by an external second-harmonic generator, delivering 25-ns (FWHM) pulses at 10 Hz repetition rate. It is well apparent that when increasing the cavity length, the slope efficiency decreases and the threshold increases. Simulations based on simple 4-level model give an appropriate modeling of the dependence of efficiency and threshold with the cavity length [70, 72].

In Fig. 4.17, the output laser energy is plotted versus cavity length for three different pump pulse durations. The details on the experiments conducted with the 0.5 and 7-ns pump sources may be found in [70]. It is clear that the long-pulse laser (25 ns) enables much more flexibility to work with long cavities, as a laser effect was observed up to $L = 15$ cm with this laser (for 50-µJ pulses at 532 nm) [72].

This opportunity to work with long cavities is the real added value of VECSOLs. It can be used to add intracavity elements that will tailor the laser output "at will". A first example is the ability to generate UV radiation that is otherwise

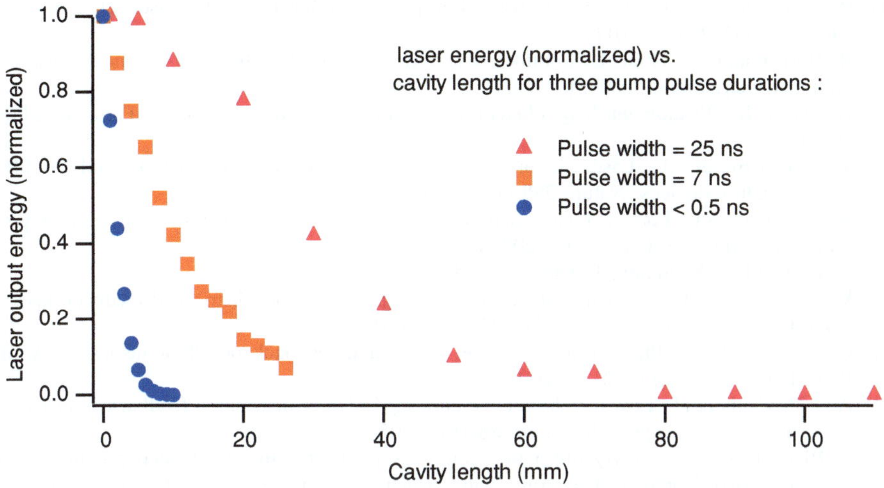

Fig. 4.17 Influence of the pump laser pulsewidth of the performance of a VECSOL. Here a rhodamine 640:PMMA laser is investigated (output coupler of 98 %, active medium thickness 17 μm)

unaccessible with organic materials, through intracavity frequency doubling [73] (see Sect. 6.1.2). It is also possible to finely control the emitted spectrum through the insertion of an intracavity Fabry-Perot etalon.

References

1. F.P. Schafer, W. Schmidt, J. Volze, Organic dye solution laser. Appl. Phys. Lett. **9**(8), 306 (1966)
2. D. Moses, High quantum efficiency luminescence from a conducting polymer in solution—a novel polymer laser-dye. Appl. Phys. Lett. **60**(26), 3215–3216 (1992)
3. A. Costela, I. García-Moreno, R. Sastre, Solid-state dye lasers, in *Tunable Laser Applications* 2nd edn., ed. by F.J. Duarte (CRC, New York, 2009), pp. 97–120
4. R.E. Hermes et al., High-efficiency pyrromethene doped solid-state dye-lasers. Appl. Phys. Lett. **63**(7), 877–879 (1993)
5. M. Faloss et al., Toward millions of laser pulses with pyrromethene- and perylene-doped xerogels. Appl. Opt. **36**(27), 6760–6763 (1997)
6. I. Garcia-Moreno et al., Materials for a reliable solid-state dye laser at the red spectral edge. Adv. Funct. Mater. **19**(16), 2547–2552 (2009)
7. M. Canva et al., Perylene-doped and pyrromethene-doped xerogel for a pulsed-laser. Appl. Opt. **34**(3), 428–431 (1995)
8. K.C. Yee, T.Y. Tou, S.W. Ng, Hot-press molded poly(methyl methacrylate) matrix for solid-state dye lasers. Appl. Opt. **37**(27), 6381–6385 (1998)
9. M.J. Cazeca et al., Epoxy matrix for solid-state dye laser applications. Appl. Opt. **36**(21), 4965–4968 (1997)
10. R. Bornemann, U. Lemmer, E. Thiel, Continuous-wave solid-state dye laser. Opt. Lett. **31**(11), 1669–1671 (2006)

11. R. Bornemann, E. Thiel, P.H. Bolívar, High-power solid-state cw dye laser. Opt. Express **19**(27), 26382–26393 (2011)
12. R. Bornemann, E. Thiel, U. Lemmer, *The Dye Laser Disk: CW Lasing from a Polymer.* Photonik international, 2007: p. 69–71
13. Okamoto, K., *Fundamentals of Optical Waveguides*, 2nd edn. (Academic Press, London, 2005)
14. V.G. Kozlov et al., Laser action in organic semiconductor waveguide and double-heterostructure devices. Nature **389**(6649), 362–364 (1997)
15. V.G. Kozlov, S.R. Forrest, Lasing action in organic semiconductor thin films. Curr. Opin. Solid State Mater. Sci. **4**(2), 203–208 (1999)
16. H. Kogelnik, C.V. Shank, J. Appl. Phys. **43**, 2327 (1972)
17. W. Holzer et al., Corrugated neat thin-film conjugated polymer distributed-feedback lasers. Appl. Phys. B-Lasers Opt. **74**(4–5), 333–342 (2002)
18. G. Heliotis et al., Blue, surface-emitting, distributed feedback polyfluorene lasers. Appl. Phys. Lett. **83**(11), 2118–2120 (2003)
19. A.E. Vasdekis et al., Low threshold edge emitting polymer distributed feedback laser based on a square lattice. Appl. Phys. Lett. **86**(16) (2005)
20. S. Riechel et al., A nearly diffraction limited surface emitting conjugated polymer laser utilizing a two-dimensional photonic band structure. Appl. Phys. Lett. **77**(15), 2310–2312 (2000)
21. N. Tsutsumi, T. Ishibashi, Organic dye lasers with distributed Bragg reflector grating and distributed feedback resonator. Opt. Express **17**(24), 21698 (2009)
22. H. Rabbani-Haghighi et al., Laser operation in nondoped thin films made of a small-molecule organic red-emitter. Appl. Phys. Lett. **95**(3), 033305 (2009)
23. A.E. Vasdekis et al., Diode pumped distributed Bragg reflector lasers based on a dye-to-polymer energy transfer blend. Opt. Express **14**(20), 9211–9216 (2006)
24. T. Riedl et al., Tunable organic thin-film laser pumped by an inorganic violet diode laser. Appl. Phys. Lett. **88**(24), 241116 (2006)
25. T. Rabe et al., Quasi-continuous-wave operation of an organic thin-film distributed feedback laser. Appl. Phys. Lett. **89**(8) (2006)
26. R. Xia et al., Low-threshold distributed-feedback lasers based on pyrene-cored starburst molecules with 1,3,6,8-attached Oligo(9,9-Dialkylfluorene) arms. Adv. Funct. Mater. **19**(17), 2844 (2009)
27. D. Schneider et al., Deep blue widely tunable organic solid-state laser based on a spirobifluorene derivative. Appl. Phys. Lett. **84**(23), 4693–4695 (2004)
28. D. Pisignano et al., Flexible organic distributed feedback structures by soft lithography. Synth. Met. **137**(1–3), 1057–1058 (2003)
29. D. Pisignano et al., Oligomer-based organic distributed feedback lasers by room-temperature nanoimprint lithography. Appl. Phys. Lett. **83**(13), 2545–2547 (2003)
30. D. Schneider et al., Organic solid-state lasers based on sexiphenyl as active chromophore. J. Appl. Phys. **98**(4) (2005)
31. S. Riechel et al., Very compact tunable solid-state laser utilizing a thin-film organic semiconductor. Opt. Lett. **26**(9), 593–595 (2001)
32. K. Baumann et al., Organic mixed-order photonic crystal lasers with ultrasmall footprint. Appl. Phys. Lett. **91**(17) (2007)
33. K. Baumann et al., Design and optical characterization of photonic crystal lasers with organic gain material. J. Opt. **12**(6) (2010)
34. G. Heliotis et al., Two-dimensional distributed feedback lasers using a broadband, red polyfluorene gain medium. J. Appl. Phys. **96**(12), 6959–6965 (2004)
35. G. Heliotis et al., Emission characteristics and performance comparison of polyfluorene lasers with one- and two-dimensional distributed feedback. Adv. Funct. Mater. **14**(1), 91–97 (2004)
36. G.A. Turnbull et al., Operating characteristics of a semiconducting polymer laser pumped by a microchip laser. Appl. Phys. Lett. **82**(3), 313–315 (2003)

37. C. Karnutsch et al., Improved organic semiconductor lasers based on a mixed-order distributed feedback resonator design. Appl. Phys. Lett. **90**(13), 131104 (2007)
38. S.V. Frolov et al., Stimulated emission in high-gain organic media. Phys. Rev. B **59**(8), R5284–R5287 (1999)
39. J. Fallert et al., Co-existence of strongly and weakly localized random laser modes. Nat. Photon **3**(5), 279–282 (2009)
40. D.S. Wiersma, Laser physics: random lasers explained? Nat. Photon **3**(5), 246–248 (2009)
41. D.S. Wiersma, The physics and applications of random lasers. Nat. Phys. **4**(5), 359–367 (2008)
42. N.M. Lawandy et al., Laser action in strongly scattering media. Nature **368**(6470), 436–438 (1994)
43. S.V. Frolov, Z.V. Vardeny, Stimulated emission in poly(p-phenylene-vinylene) films. Synth. Met. **111**, 507–508 (2000)
44. M. Anni et al., Emission properties of organic random lasers. Phys. Status Solidi (c) **1**(3), 450–453 (2004)
45. C. Vanneste, P. Sebbah, H. Cao, Lasing with resonant feedback in weakly scattering random systems. Phys. Rev. Lett. **98**(14), 143902 (2007)
46. P.W. Anderson, Absence of diffusion in certain random lattices. Phys. Rev. **109**(5), 1492–1505 (1958)
47. J. Billy et al., Direct observation of Anderson localization of matter waves in a controlled disorder. Nature **453**(7197), 891–894 (2008)
48. M. Störzer et al., Observation of the critical regime near Anderson localization of light. Phys. Rev. Lett. **96**(6), 063904 (2006)
49. D.S. Wiersma et al., Localization of light in a disordered medium. Nature **390**(6661), 671–673 (1997)
50. N. Tessler, G.J. Denton, R.H. Friend, Lasing from conjugated-polymer microcavities. Nature **382**(6593), 695–697 (1996)
51. V. Bulovic et al., Transform-limited, narrow-linewidth lasing action in organic semiconductor microcavities. Science **279**(5350), 553–555 (1998)
52. A. Schulzgen et al., Near diffraction-limited laser emission from a polymer in a high finesse planar cavity. Appl. Phys. Lett. **72**(3), 269–271 (1998)
53. L. Persano et al., Low-threshold blue-emitting monolithic polymer vertical cavity surface-emitting lasers. Appl. Phys. Lett. **89**(12) (2006)
54. M. Koschorreck et al., Dynamics of a high-Q vertical-cavity organic laser. Appl. Phys. Lett. **87**(18) (2005)
55. H. Sakata, H. Takeuchi, Diode-pumped polymeric dye lasers operating at a pump power level of 10 mW. Appl. Phys. Lett. **92**(11) (2008)
56. I.D.W. Samuel, E.B. Namdas, G.A. Turnbull, How to recognize lasing. Nat. Photonics **3**(10), 546–549 (2009)
57. B. Schutte et al., Continuously tunable laser emission from a wedge-shaped organic microcavity. Appl. Phys. Lett. **92**(16) (2008)
58. S.V. Frolov, Z.V. Vardeny, K. Yoshino, Plastic microring lasers on fibers and wires. Appl. Phys. Lett. **72**(15), 1802–1804 (1998)
59. G. Ramos-Ortiz et al., Temperature dependence of the threshold for laser emission in polymer microlasers. Appl. Phys. Lett. **77**(18), 2783 (2000)
60. R. Baktur, L.W. Pearson, J. Ballato, Theoretical determination of lasing resonances in a microring. J. Appl. Phys. **101**(4), 043102 (2007)
61. S.V. Frolov et al., Cylindrical microlasers and light emitting devices from conducting polymers. Appl. Phys. Lett. **72**(22), 2811–2813 (1998)
62. S.X. Dou et al., Polymer microring lasers with longitudinal optical pumping. Appl. Phys. Lett. **80**(2), 165–167 (2002)
63. S.V. Frolov et al., Lasing and stimulated emission in pi-conjugated polymers. IEEE J. Quantum Electron. **36**(1), 2–11 (2000)

64. M. Lebental et al., Directional emission of stadium-shaped microlasers. Phys. Rev. A, **75**(3) (2007)
65. M. Lebental et al., Inferring periodic orbits from spectra of simply shaped microlasers. Phys. Rev. A **76**, 023830 (2007)
66. M. Berggren et al., Solid-state droplet laser made from an organic blend with a conjugated polymer emitter. Adv. Mat. **9**(12), 968 (1997)
67. M. Zavelani-Rossi et al., Organic laser based on thiophene derivatives. Synth. Met. **139**(3), 901–903 (2003)
68. H. Rabbani-Haghighi et al., Highly efficient, diffraction-limited laser emission from a vertical external-cavity surface-emitting organic laser. Opt. Lett. **35**(12), 1968–1970 (2010)
69. M. Kuznetsov et al., High-power (>0.5-W CW) diode-pumped vertical-external-cavity surface-emitting semiconductor lasers with circular TEM00 beams. IEEE Photonics Technol. Lett. **9**(8), 1063–1065 (1997)
70. H. Rabbani-Haghighi et al., Analytical study and performance optimization of vertical external cavity surface-emitting organic lasers. Eur. Phys. J. Appl. Phys. **56**(3) (2011)
71. A. Tyagi et al., Photophysical characterization of pyrromethene 597 laser dye in cross-linked silicon-containing organic copolymers. Chem. Phys. **342**(1–3), 201 (2007)
72. S. Chénais et al., Laser turn-on behavior in organic vertical-external cavity surface-emitting lasers. Proc. of SPIE **8433**, 84331N (2012). doi:10.1117/12.922655
73. S. Forget et al., Tunable ultraviolet vertically-emitting organic laser. Appl. Phys. Lett. **98**(13), 131102 (2011)

Chapter 5
Novel Concepts for Organic Lasers

Abstract In this chapter, we attempt to analyze some novel research tracks in the field of organic lasers. The quest of the "organic laser diode", or electrically-pumped organic semiconductor laser, is not properly speaking a novel research axis as it has been driving the scientific community since almost the beginning of the story of OSC development. However it remains unrealized at the time of writing this book, and it is still a major scientific challenge. Many works have contributed to a better understanding of the bottlenecks preventing electrical pumping of organic lasers, they are briefly reviewed here. In the meantime, new and alternative concepts came to birth. Optically-pumped organic lasers ceased to be regarded as intermediate steps leading to a future electrically-pumped device; instead they started to be regarded as interesting devices on their own, as far as cheap pump sources, such as laser diodes or even light-emitting diodes, could be used. A fundamental challenge which remains to be solved is the realization of a true CW laser with an organic solid-state medium. By "true CW" it is meant that the laser medium is not physically moving to mimic dye circulation. Triplets are central in this problem and they have to be "managed" in some way; some interesting solutions have been proposed which are also briefly reviewed. Finally, Interesting new research fields have recently emerged using organic gain media in devices that are not classical lasers anymore, but rather devices based on quasi-particles such as exciton-polaritons (mixing of photon states and excitons) or Surface Plasmons (and Surface Plasmon Polaritons). The latter are called spasers, not all of them are organic but these devices can clearly beneficiate from the high gain provided by organic materials; they offer great promises in the realization of sub-wavelength coherent sources.

S. Forget and S. Chénais, *Organic Solid-State Lasers*,
Springer Series in Optical Sciences 175, DOI: 10.1007/978-3-642-36705-2_5,

5.1 The Organic Laser Diode

5.1.1 History of a Long Quest

Soon after an optically-pumped organic solid-state semiconductor laser was demonstrated in 1996 [1], an electrically-pumped organic laser, or "organic laser diode", became an important breakthrough worth to be targeted, potentially opening the way to a new class of extremely compact, tunable, cheap, and why-not-flexible lasers: it remains a hot topic as no convincible organic laser diode has been published at the time of writing (2012).

In 2000, Schön et al. [2] reported such an achievement in a tetracene crystal, but this paper (and others by the same author) was later retracted [3], a story which unfortunately entered the history of scientific misconduct rather than the history of science. Other claims of electrically-driven stimulated emission have been reported during the XXIst century: Yokoyama et al. [4] observed a spectral narrowing of the edge emission of an OLED as well as an intriguing superlinear dependence of the emission with the electrode length, which was attributed to the presence of stimulated emission. Tian et al. [5] observed the same features but suggested that a misalignment of the collecting optics could be the reason for the observed superlinear dependence instead of gain. In both cases, the spectral narrowing can be well accounted by a resonant leaky mode at the cut-off frequency [4]. Another intriguing claim of lasing under electrical pumping in a microcavity OLED was reported by Liu et al. [6] with a puzzling low threshold. Those recent reports, however, are probably not lasing, as shown and discussed by Samuel et al. [7]. It can be added that stimulated emission is claimed in Ref. [4] under CW excitation, which is a strong indication against lasing: CW operation under optical pumping would be actually quite a breakthrough in itself as discussed later in Sect. 5.3 (such a report by Nakanotani et al. [8] is questionable for the same reasons and is probably due to a resonant leaky mode).

5.1.2 Evidence of Extra-Losses Compared to Optical Excitation

The challenges to realize direct electrical pumping of a laser device are numerous and complex [9–11], and begin to be well understood. Quite unfortunately, the astonishing progresses made in OLED technology in the past ten years, in terms of luminance and efficiency, are not straightforwardly transferrable to organic lasers (maybe in contrast we could expect from the OLED community interesting advances in terms of device lifetime improvements or progresses in device processing techniques).

A good illustration of a breakthrough for OLEDs that is not directly useful for lasers is triplet energy harvesting through the use of phosphorescent materials [12].

This new class of materials led to a ~fourfold increase of efficiency as 75 % of the excitons created by carrier recombination are triplets. However, phosphorescence is not straightforwardly suitable for lasing as discussed in (at least attempts to observe ASE in highly-efficient phosphorescent emitters $Ir(ppy)_3$ [13] or $Btp_2Ir(acac)$ [14] have failed) since the excited state absorption band $T_1 \rightarrow T_n$ ($n > 1$, a dipole allowed transition associated with high absorption cross sections $\sigma_{T\text{-}T}$) spectrally overlaps the $T_1 \rightarrow S_0$ emission band, associated with low stimulated emission cross sections.

The key difference between an OLED and a laser is that lasing requires net gain, whereas fluorescent emission is a linear thresholdless phenomenon: because π-conjugated systems have short excited state lifetimes (~a few ns), a high exciton creation rate is required to run a laser, hence a high current density, probably much more than in a standard operating OLED, which is not easy as the mobility of OSCs is typically low.

It is instructive to seek an estimation of what would be the current density threshold of an organic laser diode if the losses were the same as in an optical pumping configuration. We start from typical data available from optically-pumped DFB lasers, where the threshold in terms of absorbed energy density is typically $E_{th} \sim \mu J/cm^2$; it can be related to the surface density n_S of singlet excitons produced at threshold $n_s^{threshold} = \frac{E_{th}}{h\nu} \sim 10^{12}\,\mathrm{cm}^{-2}$.

For the sake of simplicity we can assume a simple OLED structure in which excitons are created at the interface between an electron-transporting material and a hole-transporting material: in this case all electrons and holes recombine (or in other words every created exciton is the result of one injected electron and one injected hole, which are necessarily equal in number). At threshold only spontaneous decay (both radiative and nonradiative) causes the exciton population to decay with a fluorescence lifetime $\tau_F = \phi_F \tau_{rad}$ (see), and the corresponding current density at threshold is hence:

$$j_{threshold} = \frac{n_s^{threshold} e}{\xi \phi_F \tau_{rad}}$$

where $n_s^{threshold}$ is the surface density of singlet excitons, e the elementary charge, ξ is the fraction of singlet excitons created upon electron excitation (typically $\xi = 0.25$, see), ϕ_F is the quantum yield of fluorescence and τ_{rad} the radiative lifetime. With $\tau_{rad} \sim 1$ ns, $\phi_F \sim 1$, we get the following order of magnitude:

$$j_{threshold} \sim 1\,\mathrm{kA/cm^2}$$

In some DFB resonators exhibiting slightly lower thresholds, values down to ~100 A/cm^2 have been estimated: these predictions were for instance obtained from a DCM:Alq_3 active medium [15] sandwiched between two cladding layers of Alq_3, as well as from a PPV polymer laser [16].

But importantly and quite unfortunately, this figure or our above rough estimation have to be understood as a *lower limit*. Indeed, current densities of the order of several kA/cm^2 have been achieved in pulsed OLEDs [17] or in light-emitting field-effect transistor (LEFET) devices [18] without ever observing lasing, and values as high as 128 kA/cm^2 have been reached in a thin film of copper phtalocyanine [19].

It is clear that electrical driving brings extra-losses which are not present when the device is optically pumped, which are due to the presence of metallic electrodes, absorption and quenching of charge carriers (polarons) and triplet excitons. We briefly review these aspects below.

5.1.3 The Electrode Issue

First, the electrodes used for electrical injection are problematic. The lowest thresholds in optically-pumped devices have been obtained with waveguide lasers (such as DFB), because they enable a long interaction with the gain medium, are consequently a natural design choice for a future organic laser diode. However, the guided mode leaks outside the high-index active part and overlaps with the absorbing metallic electrodes, a feature which is amplified by the fact that all organic films tend to have similar indices of refraction, not helping optical confinement. As a result, the threshold is greatly increased, as it has been shown by optically pumping a structure with passive metallic contacts [20]. It is possible to carefully design the contact geometry so that the overlap between the optical mode and the electrodes is reduced as far as possible [21], for instance by putting the contacts in the node of a TE_2 waveguide mode as illustrated on Fig. 5.1 [22], by using thick charge-transport conducting polymers to keep the mode away from the electrode [23], or by using transparent conductive oxides for both electrodes [24, 25].

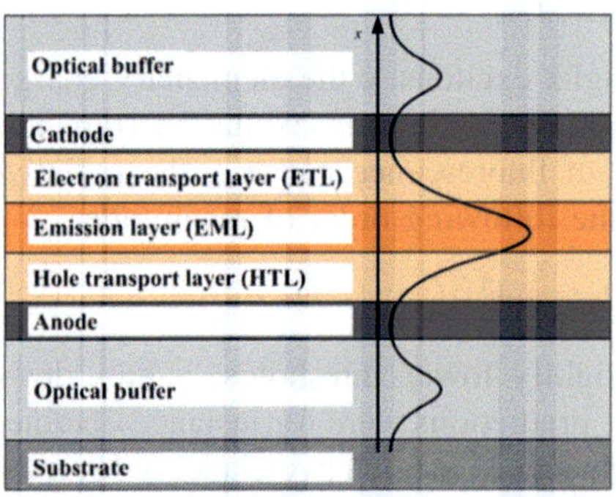

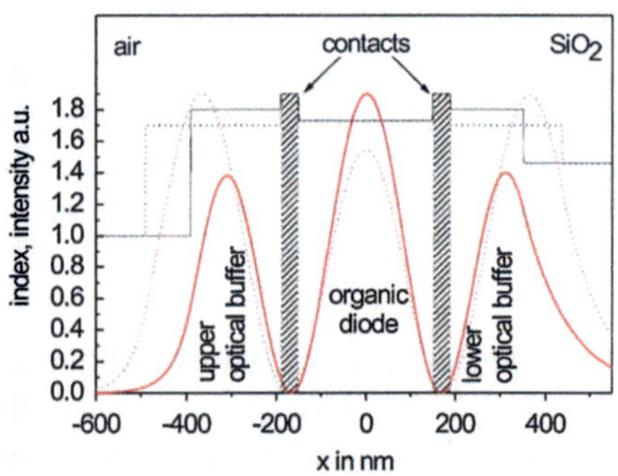

Fig. 5.1 An organic laser diode structure based on an OLED structure, designed for propagation of a TE_2-mode having a minimized overlap with the TCO electrodes. (*left* schematic illustration from [81], adapted from [22]). This design combines thin organic transport layers and a low attenuation coefficients of the waveguide

5.1.4 The Polaron Issue

The second and troublesome source of losses associated with electrical pumping is the presence of the charge carriers themselves (polarons) which have broad absorption bands overlapping the emission and can then absorb laser photons or quench singlet excitons [26]. Little quantitative information was available until recently about polaron absorption: Rabe et al. [27] made a precise and unambiguous measurement of a polaron absorption cross section in a hole-transporting spiro bifluorene (S-TAD) compound and found a low value ($\sigma < 2.6 \times 10^{-18}$ cm^2 between 560 and 660 nm), confirming the idea that polarons more certainly interact through polaron-exciton quenching [28] than through direct photon absorption.

Both the electrode and the polaron issues are in fact directly related to the low mobility of organic semiconductors. Indeed, the mobilities μ of amorphous organic semiconductors are in the range 10^{-5}–10^{-2} cm^2 V^{-1} s^{-1}, which means that achieving a high current density $j = n.e.\mu.F$ (with n the charge carrier density, e the electron elementary charge, μ the mobility and F the electric field) will mobilize a high number n of polarons. A low mobility also means a high resistivity, meaning that the OSC is mostly an insulator: the electrodes have to be set close to each other separated by a short distance d to provide a high electric field $F = V/d$ for a reasonable applied voltage. Electrically-doped regions are used in *p-i-n* OLEDs to obtain ohmic contacts, overcome the injection problems and use lower voltages, but in the context of lasers these doped regions, sources of intense light absorption and quenching, are not appealing.

A promising laser architecture in this context is the Organic Light-Emitting Field-Effect Transistor (OLEFET) [29], which shows potential for solving the electrode and the polaron problems simultaneously. The key idea is that the field-effect mobility is higher than the bulk mobility measured in a planar OLED structure [30], so that higher current densities can be obtained (values in the range of several kA/cm^2 have been reported in OLEFETs [18, 31]). This configuration also makes it possible to position the emission zone far away from the metallic contacts. Recently, optically pumped lasing was demonstrated in a LEFET structure with a DFB grating [32]. The structure and the lasing results are detailed in Fig. 5.2. Although OLEFETs are now able to outperform the equivalent OLEDs [33], no laser action has been shown yet.

5.1.5 The Triplet Exciton Issue

The presence of triplet excitons is the third and last impediment towards organic laser diodes, and it is certainly the most difficult to overcome. Triplets are more abundant than singlet excitons under electrical excitation: firstly because of the 3:1 unfavourable creation probability ratio, and secondly because their much longer lifetime

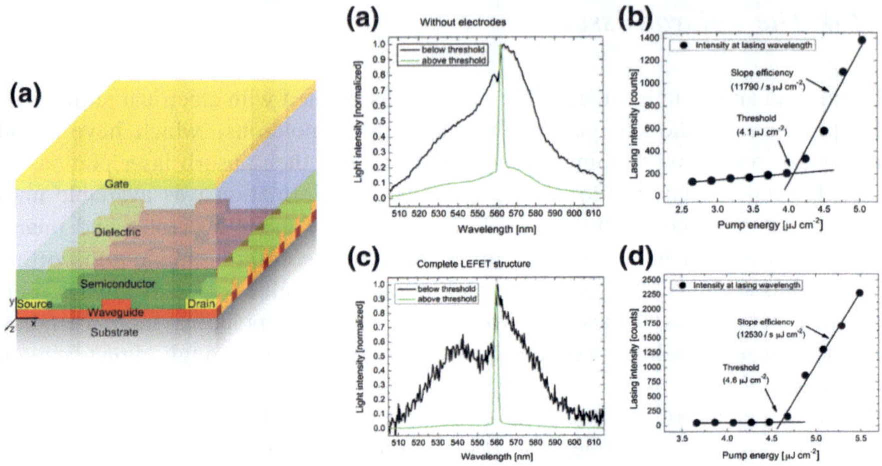

Fig. 5.2 (*Left*) a Light-emitting field-effect transistor (LEFET) structure incorporating a DFB grating for lasing applications. *Right* Spectra and laser curves of the structure without (on *top*) and with electrodes (*bottom*). Courtesy of Henning Sirringhaus. Reprinted with permission from [32]. Courtesy of H. Sirringhaus, Copyright © 2009 WILEY-VCH Verlag GmbH & Co. KGaA, Weinheim

($\sim$ms [34]) makes them piling up, at least in a long-pulse or high repetition-rate regime. This means that short pulses are always preferable [35], but even in the case of short electrical pulses there are always three times more triplets produced. In contrast, under optical pumping, triplets are only created through intersystem crossing, and are negligible when the device works under short-pulse pumping.

The negative influence of triplets on laser action has been discussed in detail in and . Triplets cause issues because they will absorb laser photons (Triplet Absorption or TA, meaning increased cavity loss, see), and/or because they will quench singlets through a Förster-type nonradiative dipole–dipole interaction referred to as Singlet–Triplet Annihilation (STA) [9] (see). These effects are not radically different in nature since they are both related to the triplet–triplet absorption cross section $\sigma_{T\text{–}T}$ and the spectral overlap between the singlet emission band and the triplet excited state absorption band $T_1 \rightarrow T_n$ (the same feature that makes organics unsuitable for CW lasing). Contrary to polaron-related phenomena, triplet issues can be studied under optical excitation with long-pulse pumping. Based on the observation of complete lasing suppression a few tens of ns after pump turn-on in Alq_3:DCM and BCzVBi:CBP lasers, Giebink et al. [9] concluded that STA between guest singlets and host triplets was the dominant mechanism, compared in particular to TA. Lehnhardt et al. [36] showed that STA and TA both play an important role in a polymer–polymer F8BT/MEHPPV guest–host system, based on experiments revealing that lasing persists for only a few nanoseconds, regardless of the pump duration. Numerical simulations [9] led to the conclusion that, even in an ideal case where polaron quenching would be totally absent, triplets hinder lasing in a guest–host system at any current density with current materials and laser architectures (see Fig. 5.3).

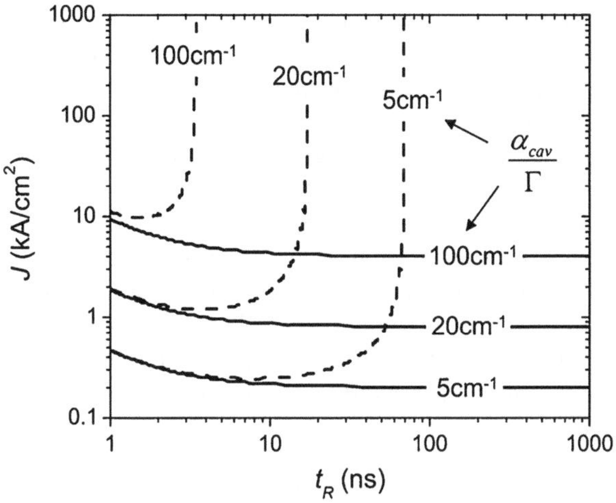

Fig. 5.3 Laser threshold criterion under electrical pumping derived from the model developed in [9]. The *lines* represent different maximum values of cavity losses (α_{cav} is the total cavity loss and Γ the confinement factor defined in Sect. 2.5.1) allowing lasing, without taking into account (*bold lines*), or with Singlet–Triplet annihilation (*dash lines*), for a given combination of t_R (the rise time of the electrical pulse) and J (current density). As an example, we can see that for a typical 100 cm^{-1} cavity losses with STA, no lasing is possible whatever the current density if the rise time is more than 4 ns. Courtesy of S.R. Forrest. Reprinted with permission from [9]. Copyright (2009) by the American Physical Society

As a conclusion, 15 years after the first demonstration of an organic semiconductor laser, achieving an organic laser diode seems as challenging as it was at the very beginning, but the bottlenecks are now much better understood. The additional losses brought by electrical excitation are now well identified and are being assessed quantitatively. For instance, Wallikewitz et al. [23] achieved a direct measurement of all the current-induced losses, by measuring the impact of a current on the threshold of an optically-pumped laser whose active medium is part of a working OLED: they revealed that a current density as low as 7 mA/cm^2 increased the threshold by 15 %. The LEFET idea, while not solving the triplet problem, suggests that real breakthroughs might be expected through a complete reformulation of the traditional concepts inherited from inorganic laser diodes.

5.2 The "Indirect Electrical Pumping" Strategy

The concept of indirect electrical pumping is to use an electrically pumped light source to (optically) pump an organic laser. This pumping light source must be cheap, compact and efficient, in contrast to expensive flashlamp-pumped Nd:YAG lasers, Nitrogen or Ti–sapphire lasers which have been the most widely used pump sources so far. This approach is then justified by the important progresses made in solid-state coherent light source technology. Very compact pulsed microchip solid-state lasers have hence been used for the pump [37–39], leading to organic sources having the size of a small shoebox. A microchip laser remains however a complex light source, composed of an inorganic infrared laser diode and a laser crystal coupled to a saturable absorber emitting infrared light in a pulsed mode, followed by frequency conversion stages to reach the UV-blue wavelengths suitable to pump most of organic lasers. A major advance in terms of reducing costs and improving compactness was triggered by the release in 1996 by the Nichia corp. of blue InGaN laser diodes. Since then, the output powers have been scaled up (driven by the need for high power DVD burner lasers for the blu-ray industry) and became suitable to directly pump organic lasers [40–42]. Going still further in simplification, Yang et al. in St Andrews demonstrated an organic semiconductor laser pumped by an inorganic incoherent light emitting diode (LED), creating a highly compact truly low-cost organic source [43].

5.2.1 Laser Diode Pumping

Using an inorganic laser diode to pump an organic laser requires two conditions: (1) the existence of reasonably efficient laser diodes emitting in the absorption range of most organic materials (namely in the UV-blue spectral area), and (2) laser materials and resonator architectures providing very low lasing thresholds to compensate the still (up to now) limited amount of power provided by the laser diodes.

These two requirements were met in the mid 2000s, leading to the first demonstration of a laser-diode-pumped solid-state organic laser in 2006 in W. Kowalsky's group in Braunschweig [41] and I. Samuel's group in St Andrews [40]. In both cases, a pulsed (In) GaN laser diode was used to pump a low threshold DFB structure. The semiconducting polymers used for the experiments were doped with dye molecules (DPAVB in PFO and coumarin in MEH-PPV respectively) to enhance the optical absorption at the blue laser diode wavelength (406 and 409 nm, respectively). In U. Lemmer's group in Karlsruhe, diode pumped organic laser with the well-known Alq_3/DCM mixture was also achieved [44]. In all cases, the lasing thresholds are comparable (a few kW per cm^2).

Sakata et al. [42] also demonstrated a diode pumped organic laser, but they used a vertical microcavity as the lasing structure. Here again, the active medium was a polymer (PVK) doped with a coumarin molecule. The lasing threshold was here higher (nearly 100 kW/cm^2).

Since those demonstrations, some progress has been made in the field of high power violet-blue laser diodes. It is now (in 2013) possible to have about 1 or 2 W at 405 or 445 nm in a single emitter, and much more when several diodes are coupled into the same bundle of optical fibers (but at the expense of a worse M^2 factor or focussability). With the foreseeable power increase and cost decrease associated with the apparition of new market routes (laser projection for example) in the next years, simple and compact diode pumped organic lasers can be an interesting alternative to expensive laser-pumped devices.

5.2.2 Light-Emitting Diode (LED) Pumping

Diode lasers remain costly, and the true low cost potential of diode-pumped organic lasers will be proved only whenever UV/blue diode lasers will be available at very low costs, which will be true if they enter mass production. There is however another class of light-emitting devices which already entered mass production, and which are already available at very low prices: these are Light-Emitting Diodes.

Light emitting diodes have been widely studied for several decades, especially since the year ~2000, driven by the explosion of LEDs in mass-production markets such as solid-state lighting and displays. In particular, blue light emitting diodes performance is increasing rapidly, and the cost per Watt of optical power is decreasing with comparable rate. However, LEDs are not the natural choice for pumping a laser: they are incoherent light sources (they cannot be focused to a diffraction-limited spot without loss unlike coherent laser sources), which makes their use as pump sources much more problematic than laser diodes. Without the possibility of tight focusing, reaching the lasing threshold is a serious challenge.

In 2008, I. Samuel's group proposed to directly put in contact a high power LED on top of a low-threshold DFB organic laser [43] (see inset in Fig. 5.4). They used a 2008 state-of-the-art LED, manufactured by Philips (Luxeon), with an output power of 620 mW over a 1.5 × 1.5 mm^2 area. After adequate pulsing of the LED (pulse duration is around 50 ns), a maximum peak intensity of 255 W/cm^2 was

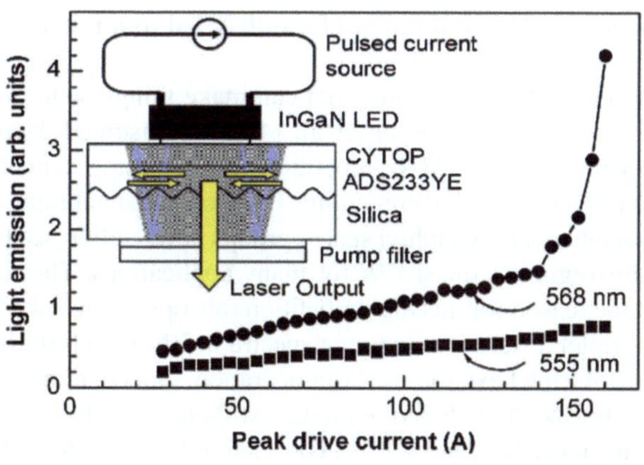

Fig. 5.4 Performance of the LED-pumped organic laser proposed by I. Samuel's group. The LED was driven with short current pulses to reach lasing threshold (details in Ref. [43]). *Inset* Scheme of the device: Pump light coming from the LED is coupled in the polymer 2nd order DFB laser through the Cytop layer. A dichroic mirror fixed to the rear surface of the substrate allows a double pass of the pump light through the polymer medium, reflecting 98 % at the pump wavelength and transmitting more than 90 % at the lasing wavelength. The high current density required to induce lasing is overcome by running the LED in a pulsed mode, with a laser-diode driver delivering pulses of duration 36 ns at 20 Hz. Courtesy of G. Turnbull. Reprinted with permission from [82]. Copyright 2008, American Institute of Physics

achieved. Without any focusing optics, this power density was enough to slightly overcome the lasing threshold of the second order DFB polymer laser.

5.3 Towards True Organic CW Lasers: Managing Triplet States

Solid-state organic lasers are pulsed devices. This aspect has been explained in detail in the introductory chapter in , where a basic analytical CW lasing condition was derived (see Sect. 2.6.1) and more advanced simulations were presented to show the dynamics of OSSLs (Sect. 2.6.3).

First experimental observations in the 60s showed that the pulse duration of a dye laser was much shorter than the pulses of the flash lamp used to pump the dye. This property originates from the piling up of triplet excitons in the organic material, which are the source of several losses (quenching processes, triplet absorption…) described above. For liquid dye lasers, a simple way to reach the CW regime is to use a liquid flow and to add triplet quenchers (typically molecular oxygen) to avoid triplet excitons accumulation.

For solid-state organic lasers, pulse durations have been for a long time limited to a few tens of nanoseconds. The quest for CW lasing motivated several attempts to mitigate the influence of triplet losses. A first proposition by Bornemann et al. (discussed previously, see Chap. 4, Sect. 4.1 and Fig. 4.1 page 109) was to mimic a liquid dye laser circulation by using a rotating disc as gain medium [45–47]. However, the mechanical fast rotation of the disk led to a relatively unstable and noisy output.

A careful choice of the organic material can make things better as far as triplet–triplet absorption is concerned: Rabe et al. [48] demonstrated a quasi-continuous organic laser operating at a 5 MHz repetition rate by using a modified poly(9,9-dioctylfluorene) (PFO) derivative containing 12 % of statistical intrachain 6,6′-(2,2′-octyloxy-1,1′-binaphtalene) binaphtyl spacer groups or BN-PFO (see Fig. 5.5). While not truly CW, this regime is quasi-CW for many applications. The same group suggested later that the reason for this high repetition rate operation is the limited overlap between triplet–triplet absorption and gain spectra in BN-PFO [49], limiting the deleterious influence of triplet excitons on laser operation. Indeed, if the same experiment is conducted with BN-PFO doped with the stilbene dye 1,4-bis(2-(4-(*N,N*-di(*p*-tolyl)amino)phenyl)vinylbenzene (DPAVB), leading to a red-shift of the gain spectrum and consequently a more pronounced overlap with the BN-PFO triplet absorption band, then the quasi-continuous regime is not possible anymore (Fig. 5.5).

Another interesting approach is "triplet management", where the unwanted triplet excitons are forced to de-excite [50, 51] on another molecule where they cannot hurt the lasing process. Zhang and Forrest [50] have published a theoretical paper predicting that a decrease in the CW threshold from 32 to 2.2 kW/cm^2, for a laser employing Alq_3/DCM as the lasing medium, if the triplets can be effectively

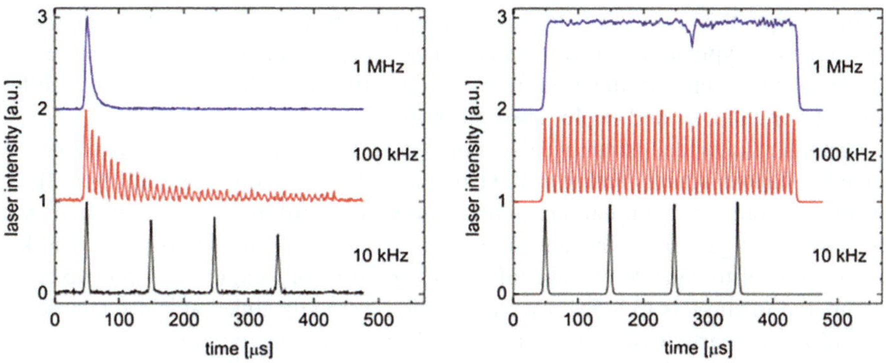

Fig. 5.5 Organic laser emission intensity in BN-PFO doped with DPAVB (*left*) and pure BN-PFO (*right*) at different repetition rates. The fast decrease observed in the *left picture* is due to triplet filling, and not due to degradation. Courtesy of T. Rabe. Reprinted from [49] with permission from Elsevier

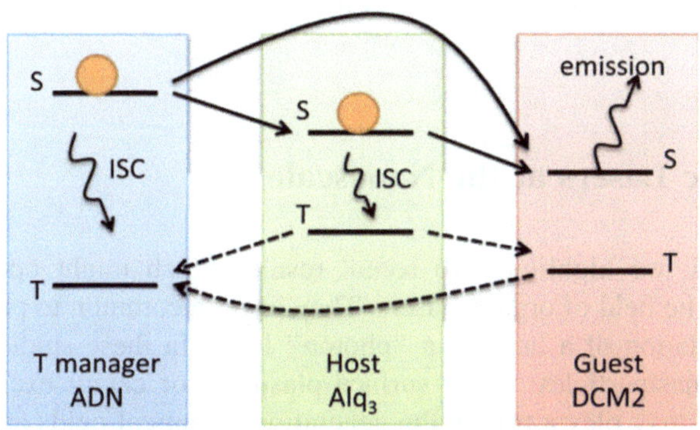

Fig. 5.6 Illustration of the different channels for exciton formation (singlet S and triplets T) and transfer between host, guest and manager molecules. Singlet are generated on both host and manager molecules, and energy is then transferred through a Förster process to the guest. Triplet generated through ISC on the host and guest molecules are collected by the triplet manager through a Dexter transfer (*dash lines*). Courtesy of S. R. Forrest. Reprinted with permission from [50]. Copyright (2011) by the American Physical Society

removed from the emissive guest. They hence show that the lasing duration can be extended to nearly 100 μs, ultimately limited by degradation of the lasing medium when a "triplet manager" molecule is blended into the gain region of an otherwise conventional distributed feedback OSL. The triplet manager has to be carefully chosen in order to have a singlet state higher in energy than the host and guest ones, but a lower triplet state (see Fig. 5.6). This way, a singlet state of the "manager" molecule, after excitation by the pump photons, can transfer its energy to the host and guest molecules. Symmetrically, the triplet states formed through ISC on the guest and host molecules transfer their energy to the triplet state of the manager. If the manager molecule triplet absorption spectrum does not overlap the gain spectra, the triplets are totally removed from the lasing system which then resembles a typical four-level system. The triplet manager, through harvesting triplet states, then facilitates radiative singlet transfer while suppressing triplet-singlet annihilation and triplet absorption as well.

Another advanced strategy for monitoring triplet decay is a clever use of metallic nanostructures acting as "plasmonic sinks" to remove the triplets [52]. By using for example metal nanoparticles as very efficient luminescence quenchers; it is possible to get rid of the unwanted long-lived states such as triplets. The key point here is to take advantage of the long diffusion length of triplet excitons compared to singlets [53–55]: by placing the plasmonic sink (typically a lattice of metallic nanoparticles) at an optimized distance of the organic molecules (for example with an effective capture radius of the sink comparable to the triplet diffusion length and much higher than the singlet diffusion length), it is possible to quench the triplets while letting the singlet states unaffected. Through a simple model, Kena-cohen et al. at Imperial College [52] predicted that for organic lasers under realistic device conditions, plasmonic sinks have the capacity to increase the achievable laser repetition rate by a factor equal to the triplet decay rate enhancement.

5.4 Organic Lasers at the Nanoscale

In this section, we highlight two recent results which might open totally new landscapes in the field of organic lasers. They have in common to put into question the very foundation of a laser as a "photon" laser. In these studies, photons are replaced by quasiparticles: either surface plasmons or cavity exciton-polaritons. While the first may play a role in the generation of optical fields at the nanoscale, the second might provide an original route to the organic laser diode problem.

5.4.1 The "Spaser" and the Advent of Nanoplasmonics

Plasmonics is a burgeoning field defined as the study of optical phenomena at the nanoscale vicinity of metal surfaces, and offers exciting perspectives of applications in nanophotonics [56], ultrafast nanoelectronics [57], or biomedical optics [58], among others.

Lasers based on the oscillation of Surface Plasmon-Polaritons instead of photons have been recently demonstrated, and referred to as "spasers" [59–61]: an excellent review article by P. Berini and I. De Leon reviews the basics and recent advances in the field of surface plasmon-polariton amplifiers and lasers [62].

Surface plasmons (SP) are collective oscillations of the conduction electrons of a metal. Surface Plasmon Polaritons (SPPs) are of particular interest: they are quasiparticles associated with electromagnetic waves coupled to free electron

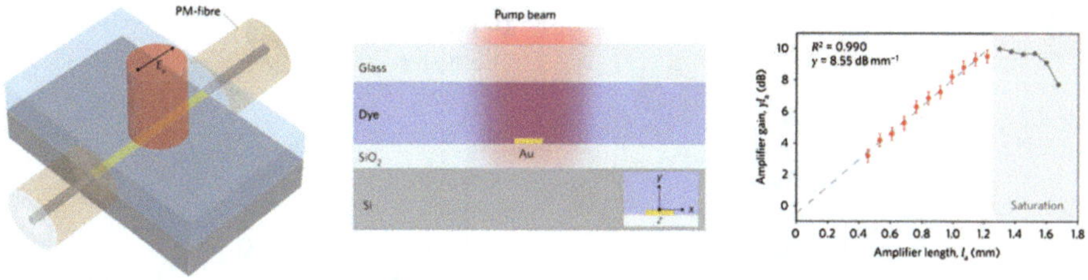

Fig. 5.7 Long range surface plasmon polariton (LRSPP) amplifier comprising a gold stripe on silica covered by a gain medium (IR-140). Courtesy of P. Berini. adapted and reprinted by permission from Macmillan Publishers Ltd: [62] (copyright 2012) and [67] (copyright 2010)

oscillations that propagate along the interface between a dielectric and a metal with a negative real part of permittivity [63]. These waves have a higher *k*-vector than in vacuum (hence, a lower wavelength) and are evanescent in both the metal and the dielectric, leading to an enhanced electromagnetic field near the interface [64]. However, because of metal absorption, SPPs cannot propagate over long distances, which is a serious impediment to build plasmonic circuitry. Many works of the last decade have then consisted in increasing the propagation lengths of SPPs by appending gain media in the dielectric. There have been several demonstrations of net gain or ASE [65–67] in plasmonic structures incorporating organic gain materials, which can be dyes in solution [66], in polymer matrices [64] or conjugated polymers [64]. Red (or infrared) dyes are especially relevant here, as a compromise between minimizing metal losses (the longer wavelength the better) and maximizing material gain. De Leon et al. [68] measured a net optical gain of 8.55 dB/mm in a Long-Range SPP (LRSPP) mode where the gain medium is a dye solution of IR-140 between silica and glass, around a thin stripe of gold (see Fig. 5.7). Similar measurements have been conducted in conjugated polymer films, such as a mixture of a PPV derivative (MDMO-PPV) with a poly (spirofluorene) polymer [64].

The study of SPP modes also brings information about organics-on-metal quenching: De Leon et al. [69] showed that only the nearest molecules (<10 nm away from the metal) are quenched through a dipole–dipole interaction to the electron/hole pairs of the metals, while most of the molecules located between 10 and 100 nm feed a Short-range SPP mode rather than a radiative photon mode.

This idea can be pushed forward to build a "Surface Plasmon Polariton laser" where the oscillating mode is a SPP: such a device was demonstrated by Oulton et al. [70] with inorganic CdS nanorod on silver. The device is characterized by a tight sub-wavelength confinement of the optical mode. Recently, Bergman and Stockman [56] proposed a different way of approaching sub-wavelength lasers with the "spaser" ("Surface Plasmon Amplification by Stimulated Emission of Radiation"): it makes use of nanoparticle (or nanolocalized) SPs, which are the quanta of the collective oscillations of free electrons in a metal nanoparticle whose dimensions are between ~ 1 nm and the metal skin depth (~ 25 nm) [60]. When such a nanoparticle is surrounded by a gain medium, the exciton on the gain

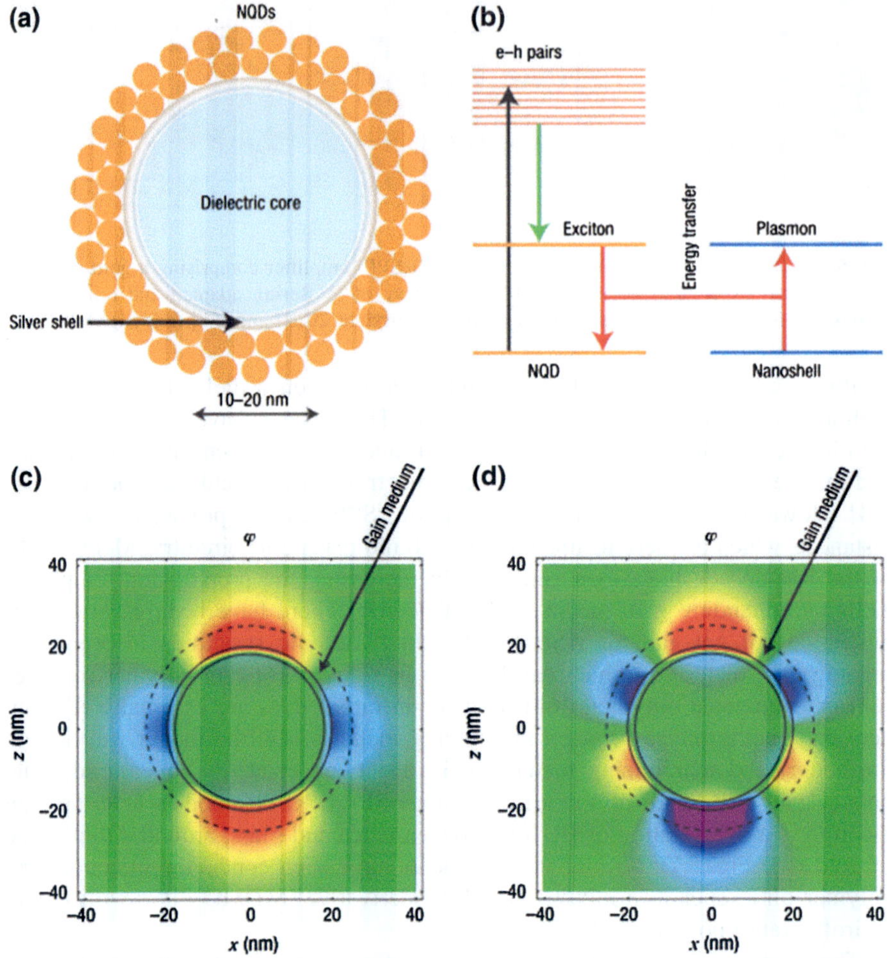

Fig. 5.8 Principle of a spaser. **a** Schematic of a spaser made from a silver nanoshell on a dielectric core (with a radius of 10–20 nm), and surrounded by two dense monolayers of nanocrystal quantum dots (NQDs). **b** Schematic of levels in a spaser. Pump radiation creates an exciton : its energy is transferred in a nonradiative way to the plasmon excitation of the metal nanoshell through resonant coupling. **c**, **d** Field amplitudes around the nanoshell in two different plasmon modes. Courtesy of M. Stockman. Reprinted by permission from Macmillan Publishers Ltd: [59] copyright (2008)

medium transfers its excitation to the Surface Plasmon modes and contributes to the accumulation of a macroscopic population of SPs in a single mode. The principle is presented in Fig. 5.8.

The first experimental demonstration of a spaser was provided in 2009 by Noginov et al. [60], who used 44-nm-in-diameter spheres composed of 14-nm-in-diameter gold cores surrounded by OG-488 dye-doped silica shells as the gain medium (see Fig. 5.9). The nanospheres were optically pumped at 488 nm and

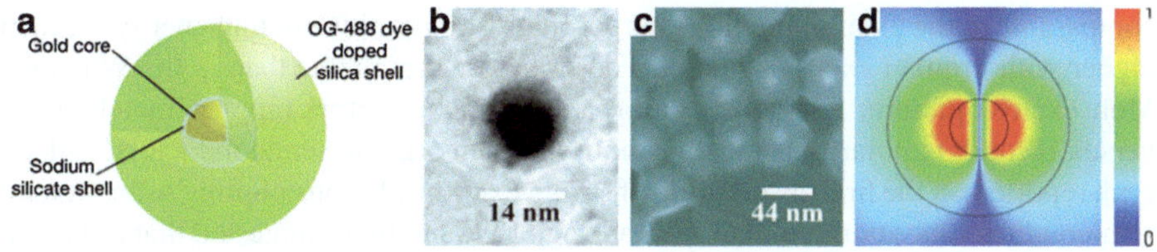

Fig. 5.9 **a** Diagram of the hybrid nanoparticle architecture (not to scale), indicating dye molecules throughout the silica shell. **b** Transmission electron microscope image of Au core. **c** Scanning electron microscope image of Au/silica/dye core–shell nanoparticles. **d** Spaser mode (in false colour), with $\lambda = 525$ nm and Q = 14.8; the *inner* and the *outer circles* represent the 14-nm core and the 44-nm shell, respectively. The field strength colour scheme is shown on the right. Courtesy of Mikhaïl Noginov. Reprinted by permission from Macmillan Publishers Ltd, Nature Ref. [60] copyright 2009

emitted a narrow line at 531 nm, which was assigned to the emission of dye molecules into a surface plasmon mode instead of a photon mode. The interest of SPASERs lies in their ability to generate electric fields oscillating at optical frequencies in *nanolocalized* areas [59]; this may be useful for microscopy (such as SNOM or fluorescence imaging with single molecule sensitivity), nanoscale lithography [56], or serve as a basis for ultrafast nanoamplifiers that have the potential to be the future building blocks of optical computers just as MOSFET transistors are the basis of today's electronics.

The nature of the gain medium in nanoplasmonic devices is, at this time of early demonstrations of physical concepts, quite unimportant, and other media such as quantum dots or inorganic semiconductors may find more useful in some applications. However the first spaser demonstrated by Noginov used organic dyes to supply gain; and it is reasonable to think that organics, as *high gain* media in the visible, added to their versatility in terms of processing techniques, will play a significant role within the next years for the development of nanoplasmonics.

5.4.2 Strong Coupling and the Organic Polariton Laser

The polariton stands at the crossroad between electromagnetic and matter waves, and its coherent state is thus somewhere between a classical photon laser [71–73] and a Bose–Einstein Condensate [74–76]. In an optical microcavity, this half-light, half-matter quasi-particle is the expression of the strong coupling between excitons and photons inside the cavity [77]. Consequently, the relevant eigenstates of the whole system are neither photons nor excitons anymore, but a linear combination of both. In the limit of a vanishing photon character and at thermal equilibrium, an exciton Bose–Einstein Condensate is obtained, whereas in the limit of a vanishing exciton character and a nonequilibrium situation the polariton laser is indistinguishable from a classical photon laser [78]. Observation of polaritons at room temperature is tricky because the excitonic part of the polariton dissociates when the thermal energy k_BT is comparable to the exciton binding energy. Interestingly, this binding energy is much higher with Frenkel excitons that best describe organic media than in the Wannier-Mott exciton picture, relevant for inorganic semiconductors (approximately 1 eV versus 10 meV) : polaritons are thus more stable in organic semiconductors. Lidzey et al. reported on the first observation of strong exciton-photon coupling in an organic semiconductor cavity more than a decade ago [79] and then suggested that polariton lasing was possible. The first demonstration of polariton lasing in an organic material was actually reported in 2010 in an anthracene single crystal [78]. The 120-nm thick crystal was sandwiched between two DBR and optically pumped by a 150-fs laser. When the threshold was reached (at 320 $\mu J/cm^2$, presumably lower than for a similar photon laser), a superlinear behavior of the output emission vs. pump power curve was observed together with a reduction of the spectral linewidth, which is said to be a clear laser signature. The spatial profile of the beam also changes to exhibit a structured TEM_{01} pattern. According to calculations, direct electrical driving of such structures is theoretically possible, opening a new exciting route to the organic laser diode.

5.4.3 The Photon Bose–Einstein Condensate

Let's mention at last an exotic example of how organic emitters can be used to build laser-like "devices", even though here a liquid dye solution was used. Thus, in optical microcavities, another exotic regime has been experimentally demonstrated for the first time in 2010 using dye molecules, namely a Bose–Einstein Condensate (BEC) of photons [80]. In a classical laser, both the state of the gain medium and the state of the light field are far removed from thermal equilibrium; in contrast a BEC of photons consists in accumulating a macroscopic population of photons, which are bosons, in the cavity ground state. The challenge was to achieve thermalization with a process that conserves the number of photons, a prerequisite for BECs, whereas usually photons in a blackbody are absorbed by the cavity walls at low temperatures. Here a BEC was realized at room temperature and thermalization achieved by absorption and re-emission processes in the dye solution. The emission of the device is composed of a nearly-monochromatic peak around 585 nm which is much alike a laser peak, with a TEM_{00} beam above critical point. Although it is clear that this demonstration may essentially serve fundamental physics purposes, the authors say it might be useful to build coherent UV sources.

References

1. F. Hide et al., Semiconducting polymers: A new class of solid-state laser materials. Science **273**(5283), 1833–1836 (1996)
2. J.H. Schon et al., An organic solid state injection laser. Science **289**(5479), 599–601 (2000)

3. Z. Bao et al., Retraction. Science **298**(5595), 961 (2002)
4. D. Yokoyama, M. Moriwake, C. Adachi, Spectrally narrow emissions at cutoff wavelength from edges of optically and electrically pumped anisotropic organic films. J. Appl. Phys. **103**(12), 123104 (2008)
5. Y. Tian et al., Spectrally narrowed edge emission from organic light-emitting diodes. Appl. Phys. Lett. **91**(14), 143504 (2007)
6. X.Y. Liu et al., Microcavity organic laser device under electrical pumping. Opt. Lett. **34**(4), 503–505 (2009)
7. I.D.W. Samuel, E.B. Namdas, G.A. Turnbull, How to recognize lasing. Nat. Photonics **3**(10), 546–549 (2009)
8. H. Nakanotani et al., Spectrally narrow emission from organic films under continuous-wave excitation. Appl. Phys. Lett. **90**(23), 231109 (2007)
9. N.C. Giebink, S.R. Forrest, Temporal response of optically pumped organic semiconductor lasers and its implication for reaching threshold under electrical excitation. Phys. Rev. B. **79**(7), 073302 (2009)
10. M.A. Baldo, R.J. Holmes, S.R. Forrest, Prospects for electrically pumped organic lasers. Phys. Rev. B **66**(3), 35321 (2002)
11. N. Tessler et al., Properties of light emitting organic materials within the context of future electrically pumped lasers. Synth. Met. **115**(1–3), 57–62 (2000)
12. Y. Sun, S.R. Forrest, High-efficiency white organic light emitting devices with three separate phosphorescent emission layers. Appl. Phys. Lett. **91,** 263503 (2007)
13. W. Holzer, A. Penzkofer, T. Tsuboi, Absorption and emission spectroscopic characterization of Ir(ppy)(3). Chem. Phys. **308**(1–2), 93–102 (2005)
14. S. Schols, *Device Architecture and Materials for Organic Light-Emitting Devices*, 1st edn. (Springer, Dordrecht, 2011)
15. V.G. Kozlov et al., Structures for organic diode lasers and optical properties of organic semiconductors under intense optical and electrical excitations. IEEE J. Quantum Electron. **36**(1), 18–26 (2000)
16. M.D. McGehee, A.J. Heeger, Semiconducting (conjugated) polymers as materials for solid-state lasers. Adv. Mater. **12**(22), 1655–1668 (2000)
17. N. Tessler, N.T. Harrison, R.H. Friend, High peak brightness polymer light-emitting diodes. Adv. Mater. **10**(1), 64 (1998)
18. E.B. Namdas et al., High performance light emitting transistors. Appl. Phys. Lett. **92**(18), 183304 (2008)
19. T. Matsushima, H. Sasabe, C. Adachi, Carrier injection and transport characteristics of copper phthalocyanine thin films under low to extremely high current densities. Appl. Phys. Lett. **88**(3), 033508 (2006)
20. P. Andrew et al., Photonic band structure and emission characteristics of a metal-backed polymeric distributed feedback laser. Appl. Phys. Lett. **81**(6), 954–956 (2002)
21. M. Reufer et al., Low-threshold polymeric distributed feedback lasers with metallic contacts. Appl. Phys. Lett. **84**(17), 3262–3264 (2004)
22. P. Gorrn et al., Loss reduction in fully contacted organic laser waveguides using TE [sub 2] modes. Appl. Phys. Lett. **91**(4), 041113 (2007)
23. B.H. Wallikewitz et al., A lasing organic light-emitting diode. Adv. Mater. **22**(4), 531 (2010)
24. S. Lattante et al., Low electrode induced optical losses in organic active single layer polyfluorene waveguides with two indium tin oxide electrodes deposited by pulsed laser deposition. Appl. Phys. Lett. **89**(3), 031108 (2006)
25. H. Yamamoto et al., Amplified spontaneous emission under optical pumping from an organic semiconductor laser structure equipped with transparent carrier injection electrodes. Appl. Phys. Lett. **84**(8), 1401–1403 (2004)
26. M.A. Baldo, R.J. Holmes, S.R. Forrest, Prospects for electrically pumped organic lasers. Phys. Rev. B **66**(3), 035321 (2002)
27. T. Rabe et al., Highly sensitive determination of the polaron-induced optical absorption of organic charge-transport materials. Phys. Rev. Lett. **102**(13), 137401 (2009)

28. C. Gartner et al., The influence of annihilation processes on the threshold current density of organic laser diodes. J. Appl. Phys. **101**(2), 023107 (2007)
29. M. Muccini, A bright future for organic field-effect transistors. Nat. Mater. **5**(8), 605 (2006)
30. C. Tanase et al., Unification of the hole transport in polymeric field-effect transistors and light-emitting diodes. Phys. Rev. Lett. **91**(21), 216601 (2003)
31. T. Takenobu et al., High current density in light-emitting transistors of organic single crystals. Phys. Rev. Lett. **100**(6), 066601 (2008)
32. M.C. Gwinner et al., Integration of a rib waveguide distributed feedback structure into a light-emitting polymer field-effect transistor. Adv. Funct. Mater. **19**(11), 1360–1370 (2009)
33. R. Capelli et al., Organic light-emitting transistors with an efficiency that outperforms the equivalent light-emitting diodes. Nat. Mater. **9**(6), 496 (2010)
34. M. Lehnhardt et al., Room temperature lifetime of triplet excitons in fluorescent host/guest systems. Org. Electron. **12**(3), 486–491 (2011)
35. S. Schols et al., Pulsed excitation of OLEDs with a remote metallic cathode. IEEE J. Quantum Electron. **46**(1), 62–67 (2010)
36. M. Lehnhardt et al., Impact of triplet absorption and triplet-singlet annihilation on the dynamics of optically pumped organic solid-state lasers. Phys. Rev. B. **81**(16), 165206 (2010)
37. H. Rabbani-Haghighi et al., Highly efficient, diffraction-limited laser emission from a vertical external-cavity surface-emitting organic laser. Opt. Lett. **35**(12), 1968–1970 (2010)
38. S. Riechel et al., Very compact tunable solid-state laser utilizing a thin-film organic semiconductor. Opt. Lett. **26**(9), 593–595 (2001)
39. G.A. Turnbull et al., Operating characteristics of a semiconducting polymer laser pumped by a microchip laser. Appl. Phys. Lett. **82**(3), 313–315 (2003)
40. A.E. Vasdekis et al., Diode pumped distributed Bragg reflector lasers based on a dye-to-polymer energy transfer blend. Opt. Express. **14**(20), 9211–9216 (2006)
41. T. Riedl et al., Tunable organic thin-film laser pumped by an inorganic violet diode laser. Appl. Phys. Lett. **88**(24), 241116 (2006)
42. H. Sakata, H. Takeuchi, Diode-pumped polymeric dye lasers operating at a pump power level of 10 mW. Appl. Phys. Lett. **92**(11), 113310 (2008)
43. Y. Yang, G.A. Turnbull, I.D.W. Samuel, Hybrid optoelectronics: A polymer laser pumped by a nitride light-emitting diode. Appl. Phys. Lett. **92**(16), 163306 (2008)
44. C. Karnutsch et al., Laser diode-pumped organic semiconductor lasers utilizing two-dimensional photonic crystal resonators. IEEE Photonics. Technol. Lett. **19**(10), 741–743 (2007)
45. R. Bornemann, U. Lemmer, E. Thiel, Continuous-wave solid-state dye laser. Opt. Lett. **31**(11), 1669 (2006)
46. R. Bornemann, E. Thiel, P.H. Bolívar, High-power solid-state cw dye laser. Opt. Express. **19**(27), 26382–26393 (2011)
47. R. Bornemann, E. Thiel, U. Lemmer, The dye laser disk: CW lasing from a polymer. Photonik. Int. pp. 69–71, (2007)
48. T. Rabe et al., Quasi-continuous-wave operation of an organic thin-film distributed feedback laser. Appl. Phys. Lett. **89**(8), 201105 (2006)
49. M. Lehnhardt et al., Spectrally separated optical gain and triplet absorption: Towards continuous wave lasing in organic thin film lasers. Org. Electron. **12**(8), 1346–1351 (2011)
50. Y. Zhang, S.R. Forrest, Existence of continuous-wave threshold for organic semiconductor lasers. Phys. Rev. B. **84**(24), 241301 (2011)
51. S. Schols et al., Triplet excitation scavenging in films of conjugated polymers. Chem. Phys. Chem. **10**(7), 1071–1076 (2009)
52. S. Kéna-Cohen et al., Plasmonic sinks for the selective removal of long-lived states. ACS Nano. **5**(12), 9958–9965 (2011)
53. N.C. Giebink, Y. Sun, S.R. Forrest, Transient analysis of triplet exciton dynamics in amorphous organic semiconductor thin films. Org. Electron. **7**(5), 375–386 (2006)
54. M. Lebental et al., Diffusion of triplet excitons in an operational organic light-emitting diode. Phys. Rev. B **79**(16), 165318 (2009)

55. R.R. Lunt et al., Exciton diffusion lengths of organic semiconductor thin films measured by spectrally resolved photoluminescence quenching. J. Appl. Phys. **105**(5), 053711 (2009)
56. M. Stockman, The spaser as a nanoscale quantum generator and ultrafast amplifier. J. Opt. **12**(2), 024004 (2010)
57. J.N. Anker et al., Biosensing with plasmonic nanosensors. Nat. Mater. **7**(6), 442 (2008)
58. H. Raether, *Surface Plasmons on Smooth and Rough Surfaces and on Gratings,* (Springer Berlin, 1988)
59. M.I. Stockman, Spasers explained. Nat. Photon. **2**(6), 327 (2008)
60. M.A. Noginov et al., Demonstration of a spaser-based nanolaser. Nature **460**(7259), 1110 (2009)
61. M.C. Gather, A rocky road to plasmonic lasers. Nat. Photon. **6**(11), 708 (2012)
62. P. Berini, I. De Leon, Surface plasmon-polariton amplifiers and lasers. Nat. Photon. **6**(1), 16–24 (2012)
63. W.L. Barnes, A. Dereux, T.W. Ebbesen, Surface plasmon subwavelength optics. Nature **424**(6950), 824 (2003)
64. M.C. Gather et al., Net optical gain in a plasmonic waveguide embedded in a fluorescent polymer. Nat. Photon. **4**(7), 457 (2010)
65. M.A. Noginov et al., Compensation of loss in propagating surface plasmon polariton by gain in adjacent dielectric medium. Opt. Express. **16**(2), 1385 (2008)
66. O. Popov, V. Lirtsman, D. Davidov, Surface plasmon excitation of amplified spontaneous emission from laser dye molecules embedded in polymer matrix. Appl. Phys. Lett. **95**(19), 191108 (2009)
67. I. De Leon, P. Berini, Amplification of long-range surface plasmons by a dipolar gain medium. Nat. Photon. **4**(6), 382 (2010)
68. I. De Leon, P. Berini, Spontaneous emission in long-range surface plasmon-polariton amplifiers. Phys. Rev. B **83**(8), 081414 (2011)
69. R.F. Oulton, Plasmon lasers at deep subwavelength scale. Nature **461**, 629 (2009)
70. D.J. Bergman, M.I. Stockman, Surface plasmon amplification by stimulated emission of radiation: quantum generation of coherent surface plasmons in nanosystems. Phys. Rev. Lett. **90**, 027402 (2003)
71. K.B. Davis et al., Bose-Einstein condensation in a gas of sodium atoms. Phys. Rev. Lett. **75**(22), 3969–3973 (1995)
72. M.H. Anderson et al., Observation of bose-einstein condensation in a dilute atomic vapor. Science **269**(5221), 198–201 (1995)
73. Q. Beaufils et al., All-optical production of chromium Bose-Einstein condensates. Phys. Rev. A **77**(6), 061601 (2008)
74. S. Christopoulos et al., Room-temperature polariton lasing in semiconductor microcavities. Phys. Rev. Lett. 98(12), 126405 (2007)
75. L.V. Butov, Solid-state physics—a polariton laser. Nature **447**(7144), 540–541 (2007)
76. H. Deng et al., Polariton lasing versus photon lasing in a semiconductor microcavity. Proc. Natl. Acad. Sci. USA **100**(26), 15318–15323 (2003)
77. A. Imamoglu et al., Nonequilibrium condensates and lasers without inversion: exciton-polariton lasers. Phys. Rev. A **53**(6), 4250–4253 (1996)
78. S. Kena-Cohen, S.R. Forrest, Room-temperature polariton lasing in an organic single-crystal microcavity. Nat. Photonics. **4**(6), 371–375 (2010)
79. D.G. Lidzey et al., Strong exciton-photon coupling in an organic semiconductor microcavity. Nature **395**(6697), 53 (1998)
80. J. Klaers et al., Bose-Einstein condensation of photons in an optical microcavity. Nature **468**(7323), 545 (2010)
81. C. Gärtner, *Organic Laser Diodes: Modelling and Simulation* (Universitätsverlag Karlsruhe, Karlsruhe, 2009)
82. Y. Yang, G.A. Turnbull, I.D.W. Samuel, Hybrid optoelectronics: a polymer laser pumped by a nitride light-emitting diode. Appl. Phys. Lett. **92**(16), 163306 (2008)

Chapter 6
Towards Applications of Organic Solid-State Lasers

Abstract While the first decade of research on organic semiconductor lasers (and more generally organic solid-state lasers) aimed at understanding the physics of such emitters and demonstrating efficient laser devices, now numerous application-oriented projects are emerging. They exploit the typical properties of organic emitters (wide tunability, easy fabrication, low thresholds and low cost) and benefit from improvements in device lifetime, output powers, beam quality or wavelength agility. We start this chapter by a brief review of recent research works that are being developed to improve the performance of organic lasers in an application-oriented view: lowering the threshold, extending the wavelength coverage, the wavelength agility (or tunability), improving the conversion efficiency, the beam quality, and the device lifetime. We then report on three major applications for organic lasers: spectroscopy, chemical sensing and short-haul communications.

6.1 Toward Real-Life Applications: The Major Issues

6.1.1 Lowering the Threshold

The laser threshold is defined as the pumping energy for which the optical gain surpasses the losses over a cavity roundtrip. Achieving a low-threshold laser system is a motivating goal, as it improves the global electrical-to-optical efficiency and lowers the amount of pump energy necessary to drive the laser. With a long term view, an electrically-pumped organic laser would be also obviously demonstrated more easily in a low-loss structure, as discussed in Sect. 5.1. For a given organic material, the lasing threshold is essentially governed by the spatial overlap between the laser mode and the pumped volume (high confinement is required), and by the quality factor of the resonator (low losses are wanted.) DFB/DBR lasers are intrinsically low-threshold lasers: typical thresholds values for

S. Forget and S. Chénais, *Organic Solid-State Lasers*,
Springer Series in Optical Sciences 175, DOI: 10.1007/978-3-642-36705-2_6,

second-order DFB lasers range from a few μJ/cm^2 [1] to tens of μJ/cm^2 [2], but results as low as 200 nJ/cm^2 [3] or 150 nJ/cm^2 [4] have been reported for small molecules blends and 40 nJ/cm^2 for conjugated polymers [5]. First order gratings have the lowest thresholds (see Sect. 3.1.1.2), but at the expense of a degraded output beam quality. A clever approach has been proposed to combine the low threshold induced by first-order gratings with the enhanced beam quality of second-order gratings: with such a mixed structure, threshold as low as 36 nJ/cm^2 were reported [6].

6.1.2 Extending the Wavelength Coverage

As seen in Sect. 2.2.5, π-conjugated systems are essentially destined to emit visible light. At longer wavelengths, nonradiative decay channels compete with fluorescence, resulting in a quantum yield of fluorescence that tends to decrease with the emitted wavelength. On the short-wavelength side, blue or UV emitters suffer from a reduced photostability, and are generally pumped with deep-UV highly energetic photons (although there has been a demonstration of 2-photon pumping of a blue polyfluorene laser [7]).

Because of the large potential interest for spectroscopy, substantial efforts have been done to look for efficient UV emitters. Silafluorenes [8] or spiro-compounds [9] are good candidates for this purpose: the lowest lasing wavelength achieved to date directly from an organic semiconductor film is 361.9 nm [10], obtained with a thermally-evaporated spiro-terphenyl film. Telecommunications, biomedical applications (for instance, deep-tissue imaging) and probably in a near future plasmonics (see Sect. 5.3.1), will keep on motivating research towards efficient deep-red or infrared gain materials and lasers. Lasing operation has been reported from 890 to 930 nm [11] in a DFB structure, or even at 970 nm in a Fabry–Perot configuration, with a commercial dye (LDS 950) doped in a fluorinated polyimide waveguide [12]. The optical gain is low in this case (14 cm^{-1}) and the lasing threshold relatively high (600 μJ/cm^2). Gain measurements have been reported at a wavelength as high as 1.3 μm in IR1051 dye [13] with a gain of 11 cm^{-1} under 1064-nm pumping.

A radically different solution to the wavelength span limit is to use the visible radiation emitted by organic systems and to convert it into UV or IR using nonlinear optics (that is, using the nonlinear dielectric properties of matter subjected to very intense fields to generate optical harmonics [14]). For a nonlinear process to be efficient, high peak intensities and good beam qualities (i.e. high brightness) must be achieved, which in practice makes this approach easier to implement with external resonators. S. Chandra et al. have obtained tunable UV radiation around 289 nm from a rotating pyrromethene dye plastic disk with external frequency doubling [15]. The same philosophy was applied by Mayer et al. [16] to obtain middle-IR radiation around 9 μm from Difference Frequency Mixing of two solid-state dye lasers emitting at 740 nm (oxazine-1) and 803 nm (Rhodamine 800).

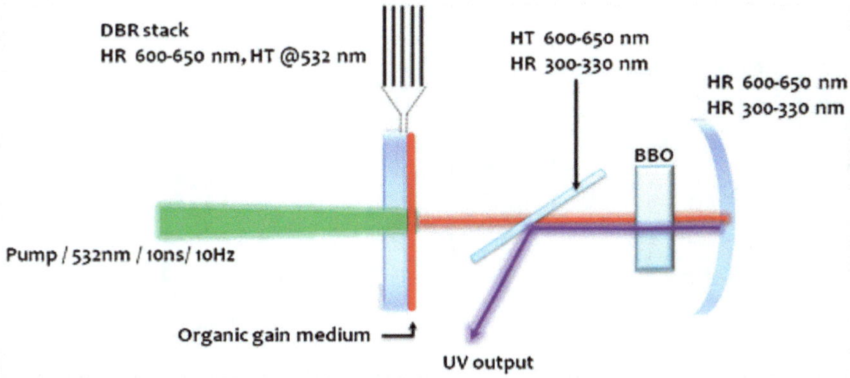

Fig. 6.1 Tunable UV organic laser based on a modified VECSOL [17]: a nonlinear crystal is inserted inside the resonator to benefit from the high peak power. A dichroic plate in the cavity allows for the extraction of the UV beam and avoids any accelerated degradation due to the interaction between UV photons and the organic layer. Reprinted with permission from [17]. Copyright 2011, American Institute of Physics

Recently, this approach was used in a thin-film organic laser with the Vertical External cavity Surface-emitting Organic Laser (VECSOL) concept presented in Sect. 4.2.3 [17]. The authors solved the low intensity problem (associated with thin-film organic lasers) by using an intracavity frequency doubling scheme, made possible by the open structure of the VECSOL. The device enables the obtention of a tunable UV organic laser at wavelengths between 309 and 322 nm, which are not accessible directly from organic emitters. The proposed device is compact (1-cm long) and emits 1 μJ of diffraction-limited tunable UV monochromatic light around 315 nm, from a Rhodamine 640:PMMA active layer spun cast onto one of the cavity mirrors (see Fig. 6.1).

6.1.3 Strategies for Advanced Wavelength Tunability

An important aspect that organic lasers must address for practical applications is the wavelength tunability or agility. In a microcavity laser, the emitted wavelength is directly related to the thickness of the active layer. Schütte et al. [18] used this property to demonstrate an Alq_3:DCM laser continuously tunable between 595 and 650 nm thanks to a wedge-shaped microcavity structure. In this structure the material was evaporated through a rotating shadow mask to a thickness varied between 180 and 1850 nm. Similarly, Rabbani et al. [19] obtained a 40-nm tunability from a VECSOL (see Sect. 4.2.3), where the spectrum is controlled by the sub-cavity formed by the active layer, upon playing on film thickness variations that arise naturally at the edges of a spun-cast film.

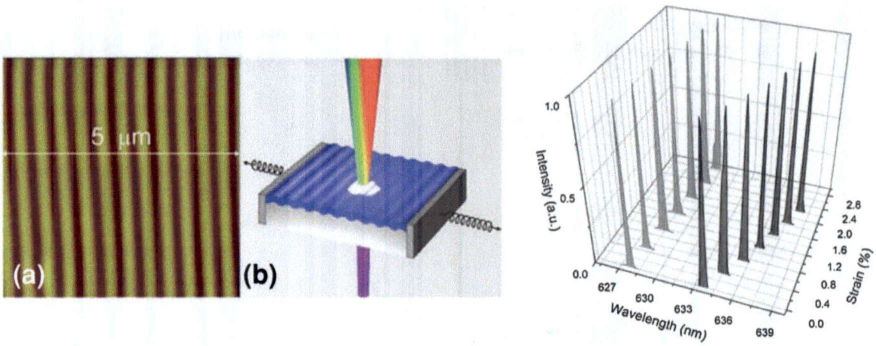

Fig. 6.2 An example of wavelength agility on a DFB laser through controlled mechanical stretching, from Ref. [22]. *Left* AFM image and artist view of the self-organized PDMS grating used as the substrate in these experiments; *right* example of wavelength coverage obtained versus applied strain in such laser (*gray* and *black lines* compare the tunability obtained with a "cast and cure" grating, obtained by replication onto PDMS (*gray*), and self-organized PDMS (*black*); the active medium is a mixture of F8BT and MEHPPV. Courtesy of P. Görnn. Reprinted with permission from [22]. Copyright © 2011 WILEY-VCH Verlag GmbH & Co. KGaA, Weinheim

DFB or DBR lasers are easily tunable upon changing the grating pitch, but the tunability is in that case achieved only by discrete steps, requiring as many gratings as desired wavelengths [20]. However, one can imagine varying the pitch in a continuous manner, for instance by using a flexible substrate subjected to a controlled mechanical stretching. Using this principle, Wenger et al. [21] demonstrated a F8BT polymer laser encapsulated in a polydimethylsiloxane (PDMS) matrix with a 20-nm tuning range around 570 nm. Görrn et al. [22] reported a lower tuning range in a F8BT/MEHPPV mixture onto PDMS with a superior resistance to stress-induced cracks using a self-organized corrugated PDMS sample, allowing for a control of the pitch (see Fig. 6.2). A flexible polymer distributed feedback (DFB) dye laser can also be combined with an electroactive dielectric elastomer actuator to enable a continuous voltage-controlled wavelength shift [23]. In DFB/DBR lasers, the emitted wavelength is also dependent on the effective refractive index of the guided mode, as shown in Chap. 4, offering another route for tuning, either by making an active nanopatterned layer with a wedged shape [24, 25] (see Fig. 6.3), or by depositing the active layer onto a wedged high index layer [26]. It has also been proposed to insert in a rhodamine-doped DBR grating some UV-sensitive molecules that are able to change their configuration (and consequently their refractive index) upon UV or visible illumination. A 14 nm tunability was thus reported, but 25 mn of illumination are needed to change the wavelength [27].

Any periodic modulation of the effective refractive index is thus potentially able to provide feedback. Besides obvious solutions such as periodic modulation of the organic layer thickness, one can also use more exotic approaches to obtain this modulation. For example, introducing laser dyes in a chiral liquid crystal allows

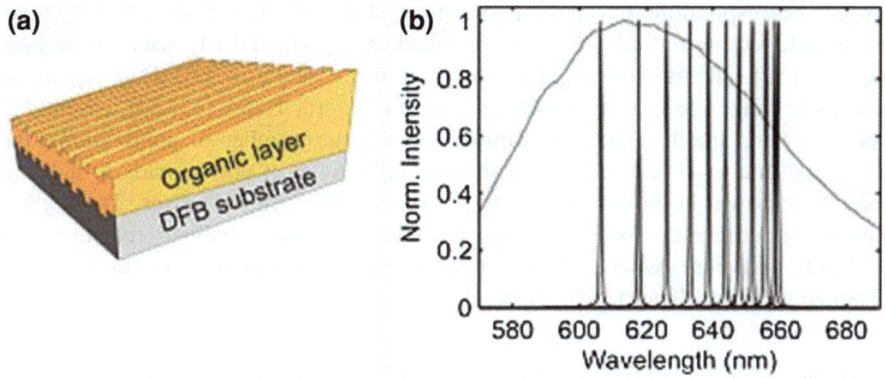

Fig. 6.3 Schematic illustration of a DFB laser with a wedge-shaped organic gain layer (*left*) and corresponding laser spectra measured at different pump spot positions on the wedge-shaped DFB laser. The photoluminescence spectrum (*gray*) of Alq_3:DCM is also drawn. Courtesy of U. Lemmer, adapted from [25] and [28]

feedback from the helicoidal shape of the crystal itself, which provide a periodic modulation of the refractive index in the direction perpendicular to the rod-like molecule orientation. The periodicity can be relatively easily changed: for example, by modifying the temperature during the curing process of the liquid crystal leads to a graded helical period, which opens the way to continuous tunability by pumping the sample at different positions. By combining such a grating pitch variation with a spatial distribution of six different dyes (some of them acting as emitters, some other as host for host–guest Förster energy transfer), full tunability is possible with a single device over the full visible spectrum [29]. It is also possible to obtain tunability by changing the pitch of the helix by a mechanical stress, by using photochromic effects (through the insertion of UV sensitive molecules, which for example flip from *cis* to *trans* configuration under UV/visible illumination), or upon applying an electric field. For more information, excellent reviews on liquid crystal lasers have been recently published [30, 31].

6.1.4 Improving the Conversion Efficiency and Output Power

Most reports on thin-film solid-state organic lasers emphasize on the spectral narrowing or threshold measurements without giving the obtained output power. The probable reason for that is the difficulty to measure such low output powers or energies with a generally very diverging, non-homogeneous, and sometimes almost poorly defined spatial beam profile. In some papers anyway this information is given and interesting slope efficiencies are obtained. For example in an early study Kozlov et al. showed a 35 % conversion efficiency (output power/input power) corresponding to a 70 % quantum slope efficiency in a Fabry–Perot waveguide configuration (see Sect. 4.2.1.1) with Alq_3:DCM_2 used as a gain medium [32]. The maximum output energy in this case was 0.9 nJ. Energies in the nanojoule range are typical for organic semiconductor lasers [33, 34], essentially due to a very small excitation volume. For a large number of applications, the nanojoule energy level could be sufficient anyway, as it corresponds to a Watt-level peak power because of the short duration of the emitted pulses. The highest conversion efficiency for a thin-film organic laser was obtained recently in a VECSOL structure, where 57 % optical-to-optical conversion efficiencies were obtained at the μJ level [35].

6.1.5 What About Beam Quality?

Beam quality (or spatial coherence) has not been a topic of major interest in early reports of organic lasers. As far as applications are concerned, beam quality is a major concern for lasers destined to be used in free space, or injected into a fiber (in a general sense, destined to be "manipulated" as a beam), but in contrast for lasers that will be part of integrated photonic circuits or for specific applications that will rely on a modification of the gain medium like sensing, beam quality is not that important. Let us remind that beam quality (see Sect. 1.4.2) is quantified by the M^2 factor which expresses the ratio of the actual beam divergence over the divergence of a diffraction-limited beam of the same size.

As previously stated (see Sects. 1.4.2 and 1.4.4), the spatial beam quality is strongly dependant on the resonator geometry [34, 36] and typical thin-film waveguide lasers may have multiple outputs and not-well-defined beams. However, it is possible to obtain a single transverse mode emission: the surface emission with a DFB structure can be nearly diffraction-limited and exhibits

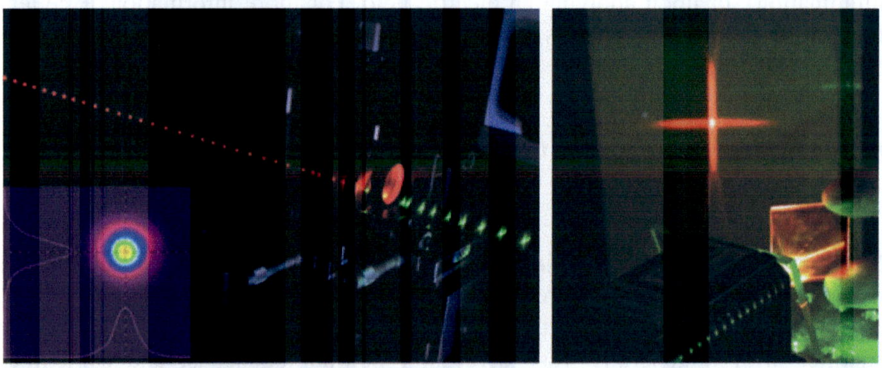

Fig. 6.4 Comparison between typical emission patterns from a VECSOL (*left*, from [19]) and from a DFB (here made of two perpendicular gratings, courtesy of I. Samuel, reprinted from [37] with permission of Elsevier, copyright 2004)

relatively low divergence [19], even if a typical fan-shaped beam instead of a desired round homogeneous gaussian beam is observed in these geometries (Fig. 6.4, *right*). For edge emission, the small aperture of the waveguide as well as the poor facet quality (especially with spin coated polymers) generally produces highly diverging and inhomogeneous beams. For vertical external-cavity surface-emitting structures, it is relatively easy to obtain TEM_{00} emission ($M^2 = 1$) because of the stability of the cavity (Fig. 6.4, *left*).

6.1.6 Enhancing the Lifetime of the Devices

All organic materials tend to degrade rapidly under intense excitation. Lifetime issues have been extensively studied in the field of OLEDs, and thanks to improvements in encapsulating techniques and materials, lifetimes up to 10 million hours have been demonstrated [38]. Because high peak intensities (and often at short wavelengths) are required to attain threshold in organic lasers, the lifetime issue happens to be more critical than in OLEDs. However, the lifetime values (when measured) reported in the literature are meaningless if the pumping conditions (pump fluence, repetition rate, ambient or oxygen-free conditions, encapsulation...) are not accurately reported, and comparison between results from various studies does generally not make sense. A typical maximum is found to be 10^5–10^7 pulses before the emission intensity decreases by a factor of two (either with microcavities [39] or DFB structures [40] under vacuum). Specific encapsulation schemes can be used to enhance the operational stability, mostly by depositing or gluing barrier layers to protect the active medium from interaction with oxygen (see [41] and references herein). It has been for example demonstrated that a careful encapsulation with an optical adhesives can strongly enhance the lifetime of a MEH-PPV DFB laser [42, 43]. The effective lifetime can thus reach hours (depending on the repetition rate), which could be enough for several applications. Another approach is to create nanocomposite materials where the fluorescent units are incorporated into a polymer host acting as an oxygen and moisture barrier [44–46]. This has also the advantage to create relatively thick self-supported "membranes", which might be disposable and cost-effective.

For bulk materials (typically dyes embedded in a host matrix), it has been shown that the nature of the matrix could have an important influence on the lifetime and stability of the laser device. For example, pyrromethene 567-based lasers can see their lifetime modified by a factor of two when embedded in various Ormosil matrices [47], to reach a typical value of several ten thousands of pulses (highly dependent of the pump intensity). Recently, an interesting study showed that by using polydimethylsiloxane (PDMS) as a host, the lifetime can be increased by a factor of 20 thanks to regeneration of the gain medium. This regeneration is possible because the dye molecules are free to move, to a certain extent, into this nanoporous matrix [48].

6.2 Spectroscopy

Among the various properties of organic lasers, their potentially wide and continuous tunability in the visible seems to make them tools of choice for spectrometry. However, various other tunable coherent sources exist on the market, and a valuable question is to determine whether organic solid-state lasers can in a near future replace those sources.

Historically, the first tunable laser source that revolutionized the field of optical spectroscopy was already organic: it was the well-known liquid dye laser, first demonstrated by Sorokin and Lankard in 1966. This kind of lasers has been reigning for decades over the field, providing for the first time a vast choice of wavelengths with reasonable power and stability to the spectroscopist.

However, and despite its tremendous qualities (high power, wide tunability, good beam quality, CW operation…), liquid dye lasers are not very popular anymore nowadays, not because of their (often unequaled) performances, but instead because of their cost, size, and handling complexity. Indeed, the circulating liquid solution, made of a toxic dye dispersed in an inflammable and toxic solvent, has to be replaced regularly to keep laser performance constant. This operation is tricky, as the concentration has to be very precisely set, and varies with the chosen dye. Additionally, a bulky pumping system is required to ensure the circulation of the dye solution. Finally, high peak powers coupled to a good beam quality is achieved at the expense of complex optical systems with pre-amplifying and amplifying systems, often complicated to align. Keeping a liquid dye laser at its best level of performance for various wavelengths requests the expertise and supervision of a trained laser technician. As a consequence, "turn-key" solid-state systems are eagerly explored.

A first very interesting solution is not a laser, but a non-linear frequency conversion system: optical parametric oscillators or OPOs. Current OPOs are able to provide coherent radiation over a huge spectral area with good conversion efficiencies (typically tens of percents) and a full-solid-state technology. The wavelength from a pump laser is converted by optical parametric generation in a nonlinear crystal set in a resonant cavity. Tuning is typically obtained by tilting the crystal or varying the temperature to change the phase-matching conditions. The main drawback of OPOs is cost (crystal growth, dichroic coatings, oven for accurate temperature control); moreover, the non-linear process is intrinsically sensitive to noise and an active control of the cavity length is often required. Finally, the spatial quality of the emitted beam is fundamentally linked to the pump beam quality (the system is not a laser, and cannot consequently be considered as a mode converter), which prevents any high-power diode-pumping scheme.

Recently, another kind of continuously tunable source based on a nonlinear process has been developed, namely supercontinuum fiber lasers. In this kind of source, the emission is not a narrow-linewidth peak that can be tuned but instead a very broad spectrum, broader than an incandescent light bulb spectrum (but with a high degree of spatial coherence). The tunability is then obtained by picking a small part of this spectrum through a set of filters or a dedicated monochromator. This broad spectrum is obtained by sending a short pulse (ps or fs) in a photonic crystal fiber. The high peak power of the pulse together with the high nonlinearity of the propagating medium leads to several complex nonlinear effects such as self-phase modulation, leading to a huge broadening of the initial laser spectrum. The main limitation of white light supercontinuum systems for spectroscopy is their limited spectral power density: as a typical value, for a 0.5 nm linewidth, commercial systems only provide a few pJ or tens of pJ per pulse.

Solid-state organic lasers are able to yield much higher energies in a narrow linewidth ($\sim\mu$J or more) and also higher peak powers ($\sim$kW or more). Organic solid-state lasers offer widely tunable laser radiation over the UV [17, 49] and the visible range and are then serious candidates for spectroscopy.

A first example was the work of Schneider et al. [49] with the demonstration of a UV tunable DBR organic laser based on spiro-linked material and emitting between 378 and 395 nm. This source was shown to be well-suited for exciting the fluorescence of various biomarkers. Deeper UV light [17] can also be generated from intracavity frequency doubled VECSOLs to extend the range of addressable biomolecules.

To demonstrate the potentiality of organic lasers for spectroscopy, Woggon et al. recently demonstrated a compact transmission spectrometer using an organic laser based on an Alq_3:DCM mixture pumped in the UV (see Fig. 6.5) [50]. By using several DFB laser with different grating pitches, continuous tunabiliy is demonstrated over 10 nm. The encapsulation in oxygen free atmosphere allows a lifetime of 500,000 pulses (before the initial intensity is divided by a factor of two). The resolution and sensitivity of the device are comparable to commercial spectrophotometers.

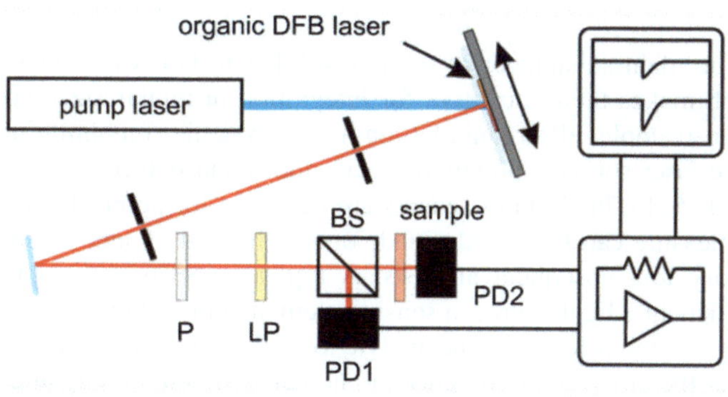

Fig. 6.5 Schematic representation of the optical setup and the detection scheme of the organic laser based transmission spectrophotometer demonstrated by Woggon et al. [50]. The organic laser is optically pumped with a pulsed UV laser. Tuning of the emission wavelength is achieved by moving the sample relatively to the pump laser beam. The laser beam is split into a probe and a reference beams which are detected using two silicon photodiodes. Courtesy of U. Lemmer, with kind permission from Springer Science+Business Media [50]

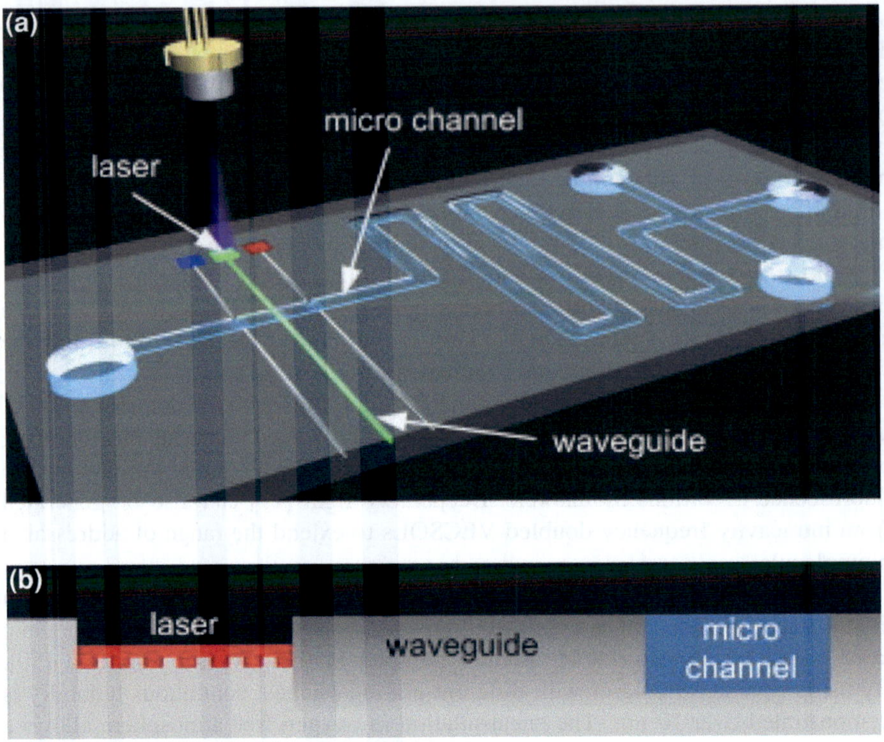

Fig. 6.6 Sketch of an uncovered, disposable lab-on-a-chip device with integrated organic lasers, deep-UV (DUV)-induced waveguides, and a microfluidic channel. **a** All organic lab-on-chip system, **b** Detection unit Courtesy of U. Lemmer, reprinted from [53] Copyright 2010, with permission from Elsevier

One of the main advantages of organic solid-state lasers is their simple fabrication which makes them attractive for integration in miniaturized spectroscopic systems. For example, all-organic lab-on-a-chip systems combining microfluidics with organic laser sources, polymeric waveguides and detectors have been demonstrated [51, 52]. This kind of sensors are desirable for point-of-care diagnostic, where the measure can be directly made on site instead of sending the sample to the laboratory for subsequent analysis. A typical spectroscopic optical sensing system consists in a light source, a sensing element and a photodetector. The most widely used scheme is to excite the analyte to be detected with a given wavelength and to detect the emitted fluorescence at another wavelength. Organic technology is able to provide a full planar sensor chip at low material cost and with easy fabrication processes, as demonstrated for example in Karlsruhe by Mappes et al. [43, 53]. A typical system is a PMMA substrate on which several Alq_3:DCM DFB lasers are engraved together with optical waveguides obtained through PMMA index modification under UV illumination (Fig. 6.6). The lasers emit at various wavelengths through a variation of the grating pitch (this tunability is crucial to be able to detect different molecules). The laser light can then easily be coupled into the organic waveguides and then directed to the optofluidic channels to excite the fluorescent molecules or markers.

Hybrid integration is also possible as organic materials can easily be integrated onto various kinds of platforms such as silicon-based photonic circuits. A promising solution to develop high performance organic optoelectronic integrated devices uses the recently demonstrated LED pumping of organic lasers to gather CMOS technology and organic semiconductors in a planar optical system. A CMOS drive circuit is here used to control a GaN LED which in-turn pumps a polymer DFB laser (as described in [54]).

6.3 Sensing

The development of high-sensitivity, reversible and stable sensors able to detect small traces of pollutant, explosives or any dangerous materials is a field of increasing interest. Nonlinear systems such as lasers can provide high response signals for a relatively small change of the input properties: from this point of view, an organic system whose properties can be slightly modified by the presence of the species to detect is a very attractive sensor. High sensitivity is linked to the nonlinear nature of the lasing process, and can be exploited through modification of the lasing threshold, slope efficiency or wavelength emission when the laser is set in an area contaminated or exposed to a given chemical compound. For example, explosives such as of 2,4,6-trinitrotoluene (TNT) are highly nitrated, and consequently electron-deficient. When they are in contact with a fluorescent polymer, an electron transfer occurs from the LUMO of the polymer and the "explosive" molecule: as a consequence, the fluorescence of the light-emitting polymer is quenched and its luminous efficiency decreases. The use of highly-nonlinear laser systems to replace simple photoluminescence quenching measurements strongly enhances the sensitivity of the method, and also speeds up the sensor response. For example Rose et al. demonstrated that trace vapours of TNT or 2,4-dinitrotoluene (DNT) can be detected by inducing fluorescence quenching of the organic polymeric gain material (here BBEHP-PPV) and consequently a modification of the lasing threshold in a planar organic laser (microdisk or DBR) [5]. In a similar way, Richardson et al. [55] and Yang et al. [56] used bisfluorene dendrimer and polyfluorene respectively in a DBR structure and showed that the slope efficiency of the laser can be strongly affected by the presence of 1,4-dinitrobenzene (DNB) (Fig. 6.7). Interestingly, the reversibility of the experiment has been demonstrated, for example through storing the sample in a nitrogen-filled box or under vacuum (see Fig. 6.8).

Another possibility is to detect a variation of the optical output spectrum when a given molecule interacts with the lasing material. Lu et al. demonstrated that the adsorption of biomolecules on a DBR organic laser can modify the effective

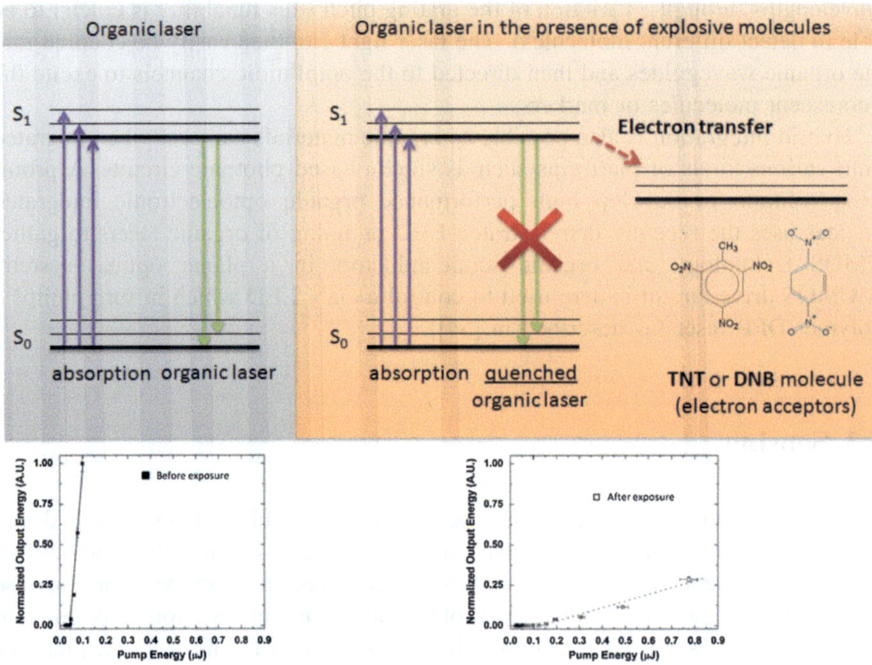

Fig. 6.7 (*Up*) illustration of the sensing principle with organic lasers: the presence of the electronegative TNT or DNB molecule quenches the laser emission, leading to a strong reduction of the slope efficiency of the organic laser. (*Bottom*) dendrimer DFB laser emission curves showing the response of the bisfluorene film before (*left*) and after (*right*) exposure to DNB. Courtesy of G. Turnbull, adapted and reprinted from [55] copyright 2009, with permission from Elsevier

refractive index of the organic layer, and that the correlated change in the emitted wavelength can be detected with good sensitivity [57].

6.4 Telecommunications

The domain of optical telecommunications is dominated by inorganic materials and technologies for sources (semiconductors), transmission (silica optical fiber) and detectors (semiconductor photodiodes). All those components are efficient, well-established and relatively inexpensive because of their fabrication on a huge scale. However, for a given set of applications where wavelength tunability or low cost fabrication processing is needed, organics can have their word. A complete survey of the use of organics for telecommunications can be found in [58].

A striking example is given by the last stage of data communication sytems (Fiber to the home or FTTH), where copper wires are responsible for the reduced

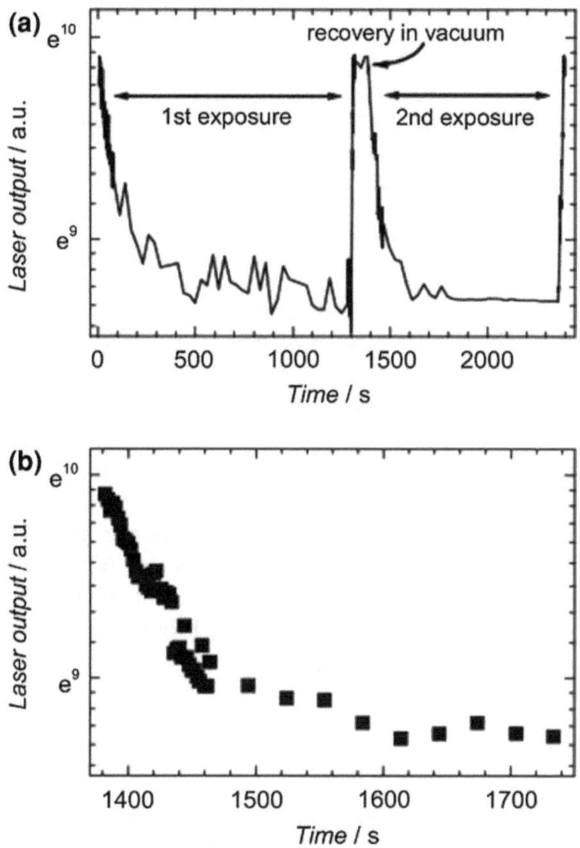

Fig. 6.8 Power characteristic of a PFO laser before and after DNT vapor exposure (*up*). Demonstration of the recovery cycle (*down*). Reprinted with permission from [56]. Courtesy of G. Turnbull, Copyright © 2010 WILEY-VCH Verlag GmbH & Co. KGaA, Weinheim

bandwidth in the local-area network and their replacement by silica fiber is not cost-effective. Despite their high attenuation coefficients (100 dB/km), polymer optical fibers (typically made of PMMA or perfluorinated polymers) are an attractive alternative for high speed and low cost connections. Their transparency window is in the orange-red part of the optical spectrum, where organic lasers are efficient and inherently integrable in planar structures or directly in optical fibers (through doping of the core by a proper dye). Organic materials can also be used as efficient high-gain (several tens of dB per cm) optical amplifiers in this spectral window [59–61]. However, organic solid-state lasers will probably not represent a viable solution in this area until lifetime issues have not been very much improved.

Another interesting application of organic amplifiers or lasers for high-speed telecommunications is the realization of an all-optical switch with fast response time. A cheap and easy to process scheme is resonant switching, where a control beam is

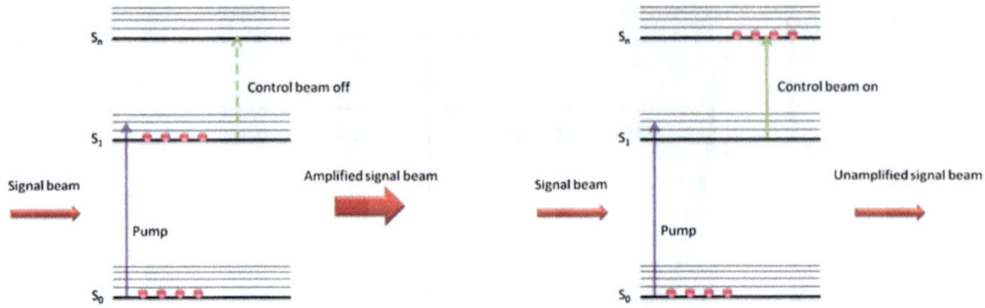

Fig. 6.9 Principle of resonant optical switching. *Left* the signal beam is amplified. *Right* the amplification is hindered because the excitons are pushed to a higher level by the control beam

used to switch on or off the amplification of a weak signal by a pump beam. Upon excitation by a pump beam, excitons are formed in the S_1 level. If a "control" laser has a wavelength that corresponds to a S_1–S_n transition (S_n corresponding to a singlet electronic level of higher energy), it will be able to promote the polymer in a higher lying state and consequently quench the optical gain (see Fig. 6.9). The amplification can then be switched on or off in a few picoseconds [60, 62]. Similar experiments were performed with lasers instead of amplifiers, showing that switching is possible with low-energy control pulses (nJ range versus μJ for the amplifier scheme) [63].

References

1. G. Tsiminis et al., Low-threshold organic laser based on an oligofluorene truxene with low optical losses. Appl. Phys. Lett. **94**(24), 243304 (2009)
2. D. Pisignano et al., Emission properties of printed organic semiconductor lasers. Opt. Lett. **30**(3), 260 (2005)
3. V.G. Kozlov et al., Structures for organic diode lasers and optical properties of organic semiconductors under intense optical and electrical excitations. IEEE J. Quantum Electron. **36**(1), 18–26 (2000)
4. R. Xia et al., Low-threshold distributed-feedback lasers based on pyrene-cored starburst molecules with 1,3,6,8-attached oligo(9,9-dialkylfluorene) arms. Adv. Funct. Mater. **19**(17), 2844 (2009)
5. A. Rose et al., Sensitivity gains in chemosensing by lasing action in organic polymers. Nature **434**(7035), 876–879 (2005)
6. C. Karnutsch et al., Improved organic semiconductor lasers based on a mixed-order distributed feedback resonator design. Appl. Phys. Lett. **90**(13), 131104 (2007)
7. G. Tsiminis et al., A two-photon pumped polyfluorene laser. Appl. Phys. Lett. **94**(25), 253304 (2009)
8. Y. Mo et al., Ultraviolet-emitting conjugated polymer poly(9,9[prime or minute]-alkyl-3,6-silafluorene) with a wide band gap of 4.0 eV. Chem. Commun. **39**, 4925 (2005)
9. N. Johansson et al., Solid-state amplified spontaneous emission in some spiro-type molecules: a new concept for the design of solid-state lasing molecules. Adv. Mater. **10**(14), 1136 (1998)
10. T. Spehr et al., Organic solid-state ultraviolet-laser based on spiro-terphenyl. Appl. Phys. Lett. **87**(16), 161103 (2005)

11. P. Del Carro et al., Near-infrared imprinted distributed feedback lasers. Appl. Phys. Lett. **89**(20), 201105 (2006)
12. S. Yuyama et al., Solid state organic laser emission at 970 nm from dye-doped fluorinated-polyimide planar waveguides. Appl. Phys. Lett. **93**(2), 023306 (2008)
13. M. Casalboni et al., 1.3 μm light amplification in dye-doped hybrid sol–gel channel waveguides. Appl. Phys. Lett. **83**(3), 416 (2003)
14. A. Yariv, *Quantum Electronics*, 3rd edn. (Wiley, New york, 1989)
15. S. Chandra et al., Tunable ultraviolet laser source based on solid-state dye laser technology and CsLiB6O10 harmonic generation. Opt. Lett. **22**(4), 209 (1997)
16. G. Mayer et al., Parametrical conversion of the frequency of organic lasers into the middle-IR range of the spectrum. Russ. Phys. J. **52**(6), 640 (2009)
17. S. Forget et al., Tunable ultraviolet vertically-emitting organic laser. Appl. Phys. Lett. **98**(13), 131102 (2011)
18. B. Schutte et al., Continuously tunable laser emission from a wedge-shaped organic microcavity. Appl. Phys. Lett. **92**(16), 163309 (2008)
19. H. Rabbani-Haghighi et al., Highly efficient, diffraction-limited laser emission from a vertical external-cavity surface-emitting organic laser. Opt. Lett. **35**(12), 19681970 (2010)
20. D. Schneider et al., Ultrawide tuning range in doped organic solid-state lasers. Appl. Phys. Lett. **85**(11), 18861888 (2004)
21. B. Wenger et al., Mechanically tunable conjugated polymer distributed feedback lasers. Appl. Phys. Lett. **97**(19), 193303 (2010)
22. P. Görrn et al., Elastically tunable self-organized organic lasers. Adv. Mater. **23**(7), 869 (2011)
23. S. Döring et al., Electrically tunable polymer DFB laser. Adv. Mater. **23**(37), 4265–4269 (2011)
24. S. Klinkhammer et al., Continuously tunable solution-processed organic semiconductor DFB lasers pumped by laser diode. Opt. Express **20**(6), 6357–6364 (2012)
25. S. Klinkhammer et al., A continuously tunable low-threshold organic semiconductor distributed feedback laser fabricated by rotating shadow mask evaporation. Appl. Phys. B Lasers Opt. **97**(4), 787 (2009)
26. M. Stroisch et al., Intermediate high index layer for laser mode tuning in organic semiconductor lasers. Opt. Express. **18**(6), 5890 (2010)
27. B.H. Wallikewitz et al., A nanoimprinted, optically tuneable organic laser. Appl. Phys. Lett. **100**(17), 173301 (2012)
28. S. Klinkhammer, T. Woggon, U. Lemmer, *Tunable Sources: Tunable Organic Semiconductor Lasers: Ready for the Market?* (Laser Focus World, 2011)
29. A. Chanishvili et al., Widely tunable ultraviolet-visible liquid crystal laser. Appl. Phys. Lett. **86**(5), 051107 (2005)
30. M. O'Neill, S.M. Kelly, Ordered materials for organic electronics and photonics. Adv. Mater. **23**(5), 566–584 (2011)
31. H. Coles, S. Morris, Liquid-crystal lasers. Nat. Photon. **4**(10), 676–685 (2010)
32. V.G. Kozlov et al., Laser action in organic semiconductor waveguide and double-heterostructure devices. Nature **389**(6649), 362–364 (1997)
33. G. Heliotis et al., Emission characteristics and performance comparison of polyfluorene lasers with one- and two-dimensional distributed feedback. Adv. Funct. Mater. **14**(1), 91–97 (2004)
34. G.A. Turnbull et al., Operating characteristics of a semiconducting polymer laser pumped by a microchip laser. Appl. Phys. Lett. **82**(3), 313–315 (2003)
35. H. Rabbani-Haghighi et al., Analytical study and performance optimization of vertical external cavity surface-emitting organic lasers. Eur. Phys. J. Appl. Phys. **56**, 34108 (2011)
36. S. Riechel et al., A nearly diffraction limited surface emitting conjugated polymer laser utilizing a two-dimensional photonic band structure. Appl. Phys. Lett. **77**(15), 2310–2312 (2000)
37. I.D.W. Samuel, G.A. Turnbull, Polymer lasers: recent advances. Mater. Today **7**(9), 28–35 (2004)

38. V.G. Kozlov et al., Study of lasing action based on Forster energy transfer in optically pumped organic semiconductor thin films. J. Appl. Phys. **84**(8), 4096–4108 (1998)
39. G. Heliotis et al., Two-dimensional distributed feedback lasers using a broadband, red polyfluorene gain medium. J. Appl. Phys. **96**(12), 6959–6965 (2004)
40. S. Richardson et al., Improved operational lifetime of semiconducting polymer lasers by encapsulation. Appl. Phys. Lett. **91**(26), 261104 (2007)
41. L. Persano et al., Rapid prototyping encapsulation for polymer light-emitting lasers. Appl. Phys. Lett. **94**, 123305 (2009)
42. B. Guilhabert et al., Amplified spontaneous emission in free-standing membranes incorporating star-shaped monodisperse Ï€-conjugated truxene oligomers. J. Opt. **12**(3), 035503 (2010)
43. C. Vannahme et al., All-polymer organic semiconductor laser chips: parallel fabrication and encapsulation. Opt. Express. **18**(24), 24881–24887 (2010)
44. T.H. Maiman, Nature **187**, 493 (1960)
45. A.E. Siegman, *Lasers* (University Science Books, Mill valley, 1986)
46. A.L. Schawlow, C.H. Townes, Infrared and optical masers. Phys. Rev. **112**(6), 1940–1949 (1958)
47. Y. Yang et al., Laser properties and photostabilities of laser dyes doped in ORMOSILs. Opt. Mater. **24**(4), 621–628 (2004)
48. H. Yoshioka et al., Fundamental characteristics of degradation-recoverable solid-state DFB polymer laser. Opt. Express **20**(4), 4690–4696 (2012)
49. D. Schneider et al., An ultraviolet organic thin-film solid-state laser for biomarker applications. Adv. Mater. **17**(1), 31–34 (2005)
50. T. Woggon, S. Klinkhammer, U. Lemmer, Compact spectroscopy system based on tunable organic semiconductor lasers. Appl. Phys. B **99**(1–2), 47–51 (2010)
51. S. Balslev et al., Lab-on-a-chip with integrated optical transducers. Lab Chip **6**(2), 213 (2006)
52. C. Vannahme et al., Plastic lab-on-a-chip for fluorescence excitation with integrated organic semiconductor lasers. Opt. Express **19**(9), 8179 (2011)
53. C. Vannahme et al., Integration of organic semiconductor lasers and single-mode passive waveguides into a PMMA substrate. Microelectron. Eng. **87**(5–8), 693–695 (2010)
54. Y. Yang, G.A. Turnbull, I.D.W. Samuel, Hybrid optoelectronics: a polymer laser pumped by a nitride light-emitting diode. Appl. Phys. Lett. **92**(16), 163306 (2008)
55. S. Richardson et al., Chemosensing of 1,4-dinitrobenzene using bisfluorene dendrimer distributed feedback lasers. Appl. Phys. Lett. **95**(6), 063305 (2009)
56. Y. Yang, G.A. Turnbull, I.D.W. Samuel, Sensitive explosive vapor detection with polyfluorene lasers. Adv. Funct. Mater. **20**(13), 2093–2097 (2010)
57. M. Lu et al., Plastic distributed feedback laser biosensor. Appl. Phys. Lett. **93**(11), 111113 (2008)
58. J. Clark, G. Lanzani, Organic photonics for communications. Nat. Photon. **4**(7), 438–446 (2010)
59. J.R. Lawrence et al., Optical amplification in a first-generation dendriticorganic semiconductor. Opt. Lett. **29**(8), 869 (2004)
60. D. Amarasinghe et al., Organic semiconductor optical amplifiers. Proc. IEEE **97**(9), 1637–1650 (2009)
61. G. Heliotis et al., Operating characteristics of a traveling-wave semiconducting polymer optical amplifier. Appl. Phys. Lett. **85**(25), 6122–6124 (2004)
62. M.M. Mroz et al., Laser action from a sugar-threaded polyrotaxane. Appl. Phys. Lett. **95**(3), 031108 (2009)
63. R. Xia et al., Wavelength conversion from silica to polymer optical fibre communication wavelengths via ultrafast optical gain switching in a distributed feedback polymer laser. Adv. Mater. **19**(22), 4054–4057 (2007)

Index

S. Forget and S. Chénais, *Organic Solid-State Lasers*,
Springer Series in Optical Sciences 175, DOI: 10.1007/978-3-642-36705-2,

MIX
Papier aus verantwortungsvollen Quellen
Paper from responsible sources
FSC® C105338

If you have any concerns about our products,
you can contact us on
ProductSafety@springernature.com

In case Publisher is established outside the EU,
the EU authorized representative is:
Springer Nature Customer Service Center GmbH
Europaplatz 3, 69115 Heidelberg, Germany

Printed by Libri Plureos GmbH
in Hamburg, Germany